AF541564

# Insect Pest Control

NIPA® GENX ELECTRONIC RESOURCES & SOLUTIONS P. LTD.
New Delhi-110 034

## About the Authors

**H. Lewin Devasahayam** after completion of his post graduate degree course from the Banaras Hindu University, Varanasi joined the Tamil Nadu Agricultural University, Coimbatore in 1960 and served the University for 30 years in various capacities and retired as Associate Professor of Plant Pathology in 1990. He has published more than 50 research articles in various leading Indian and Foreign journals. He is the author of the books entitled 'Practical manual of Entomology', 'Elements of Entomology', 'Illustrated Plant Pathology' and co-author of the books 'Crop diseases' and 'Diseases of spice crops', and all the above books have been published by NIPA, New Delhi. He is also the co-author of the book entitled 'Plant Ecology' published by M/S. Agri-Biovet Press, New Delhi.

**Dr. L. Darwin Christdhas Henry** obtained his B.Sc. Degree in Botany from the Madras University Chennai. He continued his under graduate and post graduate degree course in Agriculture at Allahabad Agricultural Institute. He joined the Faculty of Agriculture, Annamalai University He has completed his doctorate degree, specializing in the field of edible mushrooms. He has published a number of research articles in National and International Journals. He has authored the books entitled 'Crop Diseases – Identification, Treatment and Management' and 'Diseases of spice crops', both published by NIPA, New Delhi and also the book entitled 'Plant Ecology' published by M/S. Agri-Biovet Press, New Delhi. He is also the co-author of the book entitled 'Illustrated Plant Pathology' published by NIPA, New Delhi. Apart from his wide exposure to different agricultural crops, he has specialized in the cultivation of different species of edible mushrooms.

# Insect Pest Control

**H. Lewin Devasahayam**
Associate Professor (Retd.)
Tamil Nadu Agricultural University
Coimbatore, Tamil Nadu

**L. Darwin Christdhas Henry**
Associate Professor (Retd.)
Faculty of Agriculture
Annamalai University, Tamil Nadu

**NIPA® GENX ELECTRONIC RESOURCES & SOLUTIONS P. LTD.**
New Delhi-110 034

**NIPA® GENX ELECTRONIC**
**RESOURCES & SOLUTIONS P. LTD.**

101,103, Vikas Surya Plaza, CU Block
L.S.C. Market, Pitam Pura, New Delhi-110 034
Ph : +91-11-43860225, Mob.: +91 9717133558, 9540816132
E-mail: newindiapublishingagency@gmail.com
Website: www.nipaersources.com



Print ISBN: 978-93-58872-62-0

ebook ISBN: 978-93-58872-17-0



Composed and Designed by NIPA®.

# Preface

According to available evidences, insects came into existence on this planet Earth several millions of years before man appeared on this Earth. To satisfy his basic needs of food, clothing and shelter, man started raising different crops. In the mean while, some of the insects had deterred from their original feeding habits and started feeding and destroying the crops grown by man. As such, man considered such insects as his prime enemy and tried to destroy them by various means. From time immemorial, man had used products obtained from plant origin as well as from animal origin for the control of harmful insects. Side by side some of the inorganic chemical compounds were also used to destroy such insects.

After the invention of the first synthetic pesticide Dichloro diphenyl trichloro ethane (DDT) in 1939, several other Organo chlorine, Organo phosphorus, Carbamate, Pyrethroid and many other pesticides were introduced. Most of these pesticides were found to be non-specific, highly toxic and the toxic residues persisted in the applied plant parts for a considerably long time. Further, most of these pesticides were not compatible with other plant protection chemicals. Because of their high mammalian toxicity many of these supposed to be old generation pesticides were banned for plant protection use. Subsequently several new pesticides were introduced for pest control. These new generation pesticides, although recommended for the control of specific insect pests, are capable of controlling many other pests besides the target pest. Most of these pesticides although highly toxic to insect pests have less mammalian toxicity and less residual toxicity. They possess quick knock down effect on flying insect pests. Most of these pesticides are compatible with other plant protection chemicals. Of late many mixed insecticide formulations and insecticide - fungicidal formulations have been introduced. These mixed formulations can control many insect pests or insect pests and fungal parasites simultaneously

In pace with the introduction of new pesticides, several plant protection appliances have also been introduced to apply plant protection chemicals systematically and uniformly over the crop or on the soil. Many kinds of hand operated and power operated sprayers, dusters, granule applicators, soil injecting guns, flame throwers etc. have been introduced. To cover extensive area of crops infested with epidemic pests such as, rice brown plant hopper, groundnut red hairy caterpillar etc. mass ground spraying or aerial spraying is taken up to control the pest at its most vulnerable growth stage within a short period of time.

This book has been written based on the syllabus of Under Graduate students of Agriculture, but we hope that this book will be of immense use to Post Graduate students of Entomology, to those who are interested in the field of Entomology, as well as to the officials working in the Department of Agriculture, especially in the field of Plant Protection. The text is also substantiated with many hand-drawn figures by the authors. For compilation of the book, informations have been collected from several books brought out by both Indian and International authors, as well as from many other sources and we are very much indebted to the authors concerned.

The authors are thankful to their colleagues and friends for their help and encouragement for successfully completing the book. The authors also express their sincere thanks to M/S. New India Publishing Agency, New Delhi for the excellent manner in which the book has been brought out.

**Authors**

# Contents

# List of Figures

# List of Appendices

# 1

# Insect Pest

## Pests

The English name **'Pest'** is derived from the Latin word **'Pestis'**, which means causing destruction. Pests are defined as 'those organisms, which compete with man in his food supply, damage his possessions and attack himself'. They are of vital concern to mankind in their everyday life, as they cause considerable damage directly as well as indirectly to crops, domestic animals, birds and to man himself. They are also responsible for causing or transmitting many serious diseases. Pests include thousands of invertebrates such as, insects, nematodes, mites, ticks, snails and slugs, crabs etc. and vertebrates such as, rats, squirrels, jackals, pigs and boars, porcupines, rabbits, deers, bats, some species of birds etc., which are destructive to plants and livestock. Several microorganisms such as, fungi, bacteria, viruses, mycoplasma, phytoplasma etc. are also pests in nature. Weeds and phanerogamic parasites are also pests in the true sense.

## Insect pests

**'Insects'** are predominantly feeders of plants and different plant parts. There is hardly any plant or plant part, which does not harbor any insect of one type or another. Insect pests are found to cause direct damage to crops by feeding on the foliage, roots, rhizomes, tubers, stems, flowers, unripe fruits, grains or seeds during their growth. Besides food crops, fibre, plantation and other economically important plants such as, flowering and ornamental plants are also equally subjected to attack by various insect pests. In addition to causing such direct damage by feeding on plant tissues or sucking and feeding on plant sap, these pests are responsible for spreading many virus and mycoplasma diseases in crops such as, rice, potato, tobacco, tomato, sugarcane, leguminous crops, lady's finger, egg plant etc., which ultimately result in considerable reduction in yield. Insect pests also cause considerable damage to stored grains and stored products of various kinds. Besides feeding on such materials, they cause deterioration in quality of the products. Further, a number of insect pests are parasitic on man, domestic animals and birds. They suck and feed on the blood of such hosts.

They are also responsible for causing or spreading a number of diseases. An insect, which multiplies beyond a certain level and causes damage beyond a certain limit attains the status of an **'insect pest'**. In general, an insect, which infests a particular crop and causes an economic loss of 5.0 per cent or more is considered to be an insect pest.

Economic classification of insects is given in **Appendix - 1**

**Insect pests causing direct damage to crops and crop produce**

Almost every plant or plant part is attacked by some insect pest or other. The damage caused by the pests and the symptoms manifested depend upon the type of mouthparts and the nature of feeding of the insects.

i) **External feeders - Biting pests**

These pests have biting and chewing type of mouthparts. These pests remain outside the plants, bite and feed on the various plant parts. Representatives of this group are locusts, grasshoppers, beetles and weevils and larval stages of several moths and butterflies, which feed on the foliage and other plant parts

Most of the leaf-feeders eat the leaves completely and defoliate the entire plant and they are generally known as **'defoliators'** (e.g.) Groundnut red hairy caterpillar *(Amsacta albistriga),* Gram cut worm *(Agrotis ipsilon),* Castor semilooper *(Achaea janata),* Teak defoliator *(Hyblaea puera),* Locusts *(Schistocerca gregaria)* etc.

Some of the leaf feeders notch the edges of leaves and feed on the tissues (e.g.) Rice grasshopper *(Hieroglyphus banian),* Sugarcane grasshopper *(Oxya* species*),* Ash weevils feeding on the leaves of plants such as, egg plant, cotton, drum stick etc. *(Myllocerus* species*)* etc.

Some others bite small, fairly round holes in the leaves and feed on the tissues (e.g.) Red pumpkin beetle *(Raphidopalpa foevicollis)*

A few leaf-feeders scrape and feed on the chlorophyll portion of the leaves leading to skeletonization of the leaves (e.g.) Grubs and adults of lady bird beetle *(Henosepilachna vigintioctopunctata)* infesting the leaves of egg plant, bitter gourd etc.; Larvae of diamond-back moth *(Plutella maculipennis)* on cabbage and cauliflower; Rice hispa adult beetles *(Dicladispa armigera)* on rice plants.

Some bite large, irregular holes in the leaves and feed on the tissues (e.g.) Cabbage semilooper *(Trichoplusia ni)* on cabbage; Tobacco caterpillar *(Spodopera litura)* on tobacco.

A few others cut and feed on the tender growing shoots and branches (e.g.) Grapevine flea beetle *(Scelodonta strigicollis)*

The leaf rollers, roll or fold the leaves, remain inside the rolls or folds, scrape and feed on the leaf tissues (e.g.) Cotton leaf roller *(Sylepta derogata);* Gingelly leaf and shoot webber *(Antigastra catalaunalis);* Mango leaf twister weevil *(Apoderus tranquebaricus)*; Rice leaf folder *(Gnaphalocrocis medinalis)*

Some construct protective coverings, cases or bags live inside and feed on the leaf tissues (e.g.) Rice case worm *(Nymphula depunctalis)*, Coconut bag worm *(Monatha alpipes)*

The bark borers make long, winding tunnels on the surface of tree trunks with silken thread, frass and bark tissues, remain inside, scrape and feed on the bark and stem tissues (e.g.) Bark borer of cashew, mango, drum stick, curry leaf plant etc. *(Indarbela tetraonis)*

Some insects cut the germinating seedlings at the ground level and eat them (e.g.) Cotton ground weevil *(Attactogaster finitimus)* infesting castor seedlings.

A few insects bite and feed on the flower buds and flowers (e.g.) Blister beetle *(Lytta tenuicollis)* feeds on the flowers of red gram, sunnhemp etc.; Rose flower chafer beetle *(Oxycetonia versicolor)* cuts and feeds on the flowers; Larvae of the red gram spotted pod borer *(Maruca testulalis)* web the flower buds and flowers together with fine silken thread, live inside and feed on the floral parts; Jasmine bud worm *(Hendecasis duplifascialis)* bores into the jasmine flower buds and feeds on the inner parts.

Some insects gnaw and cut the peduncles of the inflorescence (e.g.) grasshoppers *(Hieroglyphus banian)* attacking rice ear heads.

The developing grains in the standing crops are sometimes eaten by a few pests (e.g.) American cotton boll worm *(Helicoverpa armigera)* feeds on the developing grains of sorghum, finger millet, sunflower etc. in the field.

**ii) External feeders - Sucking pests**

These pests have piercing and sucking type of mouthparts. They remain outside the plants, pierce and suck the sap from leaves, tender branches, floral parts, young pods, immature seeds at milk stage, fruits etc. The representatives of this group are chiefly insects belonging to Hemiptera, some species of Lepidoptera, Diptera etc. Continuous sucking of sap by numerous pests from plant parts, drastically affects the growth of the plant or plant parts resulting in marked reduction in yield and quality of the produce. The sucking pests, while sucking the sap, inject saliva into the tissues to dilute the sap, so as to facilitate sucking of the sap. The saliva of several such insects contain some toxic substances, which may aggravate the extent of damage. Besides this, some sucking insects transmit certain diseases caused by viruses, mycoplasma etc.

Many sucking pests infest the foliage, which may lead to chlorosis and the leaves become yellow and ultimately get blighted (e.g.) Lab lab aphid *(Aphis craccivora)* infesting lab lab, groundnut, glyricidia etc.; Rose thrips *(Rhpiphorothrips cruentatus)* infesting rose, grapevine etc.

Due to the infestation pale yellow spots or patches may develop at the infested area (e.g.) castor white fly *(Trialeurodes ricini)*; coconut scale insect *(Aspidiotus destructor)* etc.

The infested leaves turn silvery or whitish in color. Because the pests suck sap from tissues below the epidermal layer, such color changes occur (e.g.) Onion thrips *(Thrips tabaci)*; Groundnut thrips *(Scirtothrips dorsalis)*; Banana lace wing bug *(Stephonitis typicus)*, which causes white feeding spots on leaves of banana and coconut.

The tips of infested leaves twist and curl and dry, generally called as **'tip drying'** (e.g.) rice thrips *(Stenchaetothrips biformis)*

Leaves of infected plants become brown and blighted and the attacked plants turn brown and present a burnt up appearance in patches in the infested field, which is termed **'hopper burn'** (e.g.) rice brown plant hopper *(Nilaparvata lugens)*; cotton and castor leaf hopper *(Amrasca biguttula biguttula)*

The infested areas, especially the tender shoots and young leaves show puckering, wrinkling and curling. Many species of aphids, thrips and leaf hoppers produce such symptoms (e.g.) Chilli thrips *(Scirtothrips dorsalis)*

In the infested plants, the young leaves become very much stunted and malformed and the young shoots present a bunchy appearance (e.g.) Mealy bugs on gliricidia and crotons *(Phenacoccus solenopsis)*

In the infested plants the flower buds, flowers, and young immature fruits are shed prematurely (e.g.) Apple San Jose scale insect *(Quadraspidiotus perniciosus)*; Rose scale insect *(Aonidiella aurantii)*

Infested fruits rot and drop off prematurely (e.g.) Citrus fruit sucking moths *(Othreis fullonica, O. materna)*

**iii) Internal feeders**

The internal feeders remain inside the plant parts, feed on the inner tissues and cause severe damage. These feeders most often have very strong biting mouthparts. The entire life cycle or one or two growth stages may be spent inside the plant parts. The female adult lays eggs on the plant parts or inserts the eggs into the plant tissue by means of its ovipositor. The larvae, which hatch out of the eggs, bore into the plant parts, feed on the inner tissues and grow. They include the stem borers, flower, fruit, pod and grain borers, leaf miners and gall forming insect pests.

a) **Stem borers.** The larvae bore into the tender shoots and feed on the inner tissues causing wilting and drying of young shoots. In some cases, the branches and even the entire infested plant may dry and eventually die (e.g.) Cotton spotted boll worms *(Earias insulana, E. fabia)*; Egg plant shoot and fruit borer *(Leucinodes orbonalis)*; Mango stem borer *(Batocera rufomaculata)*

   The larvae bore into the shoots, feed on the inner tissues and in this process, the growing shoots are severed from inside, causing the emerging shoots to rot and dry, which is commonly known as **'dead hearts'** (e.g.) Rice stem borer *(Scirpophaga incertulas)*; Sorghum shoot borer *(Chilo zonellus)*; Sugarcane early shoot borer *(Chilo infuscatellus)*; Finger millet pink borer *(Sesamia inferens)*

   The larvae directly burrow into the stem and feed on the inner issues (e.g.) Sugarcane interrnode borer *(Chilo sacchariphagus indicus)*. The adult beetles bore through the shoot portion and the adult and the grubs feed on the inner tissues (e.g.) Coconut red palm weevil *(Rhynchophorus ferrugeneus)*

   The adult insects bore through the tender shoot portion and feed on the inner tissues (e.g.) Coconut rhinoceros beetle *(Oryctes rhinoceros)*

b) **Borers of various plant parts other than the stem.** The larvae bore into the flower buds and feed on the inner parts, as a result the buds drop down (e.g.) Drumstick bud worm *(Noorda moringae)*; Jasmine bud worm *(Hendecasis duplifascialis)*

   The caterpillars bore into the bolls, pods or capsules and feed on the inner content (e.g.) cotton spotted boll worms*(Earias insulana, E. vitella)*; Cotton pink boll worm *(Platyedra gossypiella)*; Egg plant fruit borer *(Leucinodes orbonalis)*; Castor capsule borer *(Dichocrosis punctiferalis)*; Red gram pod borer *(Exelastis atomosa)*

   The larvae bore through the fleshy part of the fruit and into the nut and feed on the kernel (e.g.) Mango nut weevil *(Sternochetus mangiferae)*

   The grubs and beetles bore into the seeds and feed on the inner contents (e.g.) Pulse beetle *(Pachymerus chinensis)*

c) **Leaf miners**. The larvae mine into the leaves and feed on the tissues in-between the epidermal layers (e.g.) Citrus leaf miner *(Phyllocnistis citrella)*, Rice hispa *(Dicladispa armigera)*

d) **Gall forming insects**. The larvae and adults of some insects attack the plant parts and as a result, irregular shaped galls of various sizes or blisters are formed. The young stages of these insects complete their life cycle inside such galls. The galls provide food for the insects, at the same time afford protection to them. Some insect species belonging to Cecidomyidae, Aphididae, Aleyrodidae and Psillidae and a few others belonging to the Order - Thysanoptera produce galls in different plant parts.

The maggots bore into the shoots and as a result a long, tusk-like gall, generally called **'silver shoot'** is produced from the leaf sheath of the shoot in which the early stages of the life cycle of the pest are passed (e.g.) Rice gall fly *(Pachydiplosis oryzae)*

The grubs bore into the stem region and produce galls of different sizes and shapes in the stem and live within the galls (e.g.) Cotton stem weevil *(Pempherulus affinis)*; Bitter gourd stem fly (*Lasioptera falcata)*

The maggots produce galls in the earheads of sorghum (e.g.) Sorghum gall fly *(Stenadiplosis sorghicola)*. The maggots bore into the developing capsules and produce galls in which they live (e.g.) Sesame gall fly *(Asphondylia sesami)*

**iv) Soil - inhabiting pests**

Some soil - inhabiting pests cause damage by biting and cutting the roots or boring into the roots and feeding on the tissues or by sucking the sap from the roots or by producing galls in the root region. Generally the larval stage alone causes such damages and all the growth stages may not be found in the soil.

The grubs of the beetle remain in the soil, bite, cut and feed on the rootlets of finger millet, sorghum, sugarcane etc. (e.g.) Chafer beetle, the grub of which is commonly known as white grub *(Holotrichia consanguinea)*

The grubs bore into the roots and feed on the inner tissues of sunnhemp plants (e.g.) Flea beetle *(Longitarsus belgamensis)*

The grubs feed externally on the roots and developing tubers (e.g.) White grubs of the chafer beetle *(Holotrichia conferta)* on potato.

The nymphs and adults remain outside on the roots in large numbers, suck and feed on the sap from the roots (e.g.) Finger millet root aphids *(Tetraneura nigriabdominalis)*

The nymphs and adults suck sap from the roots and also cause galls in the roots (e.g.) Apple wooly aphid *(Eriosoma lanigera)*

The caterpillars bore into the tubers and feed on the inner contents (e.g.) Potato tuber moth *(Gnorimoschema operculella)*; Grubs of the sweet potato weevil *(Cylas formicarius)*

The nymphs and adults bore into the pods and feed on the kernels (e.g.) groundnut earwig *(Euborellia stali)*

The insects bite and feed on the roots, stem and bark (e.g.) Termites attacking sugarcane, coconut, neem etc. *(Odontotermes obesus)*

**v) Storage pests**

Several pests, mostly belonging to Coleoptera and Lepidoptera attack stored grains, seeds, stored products, food products etc., resulting in considerable loss. Stored grains and harvested products may be infested before harvest in the fields or after harvest in the godowns, store houses etc.

The larvae bore into the tubers and feed on the inner contents (e.g.) potato tuber moth *(Gnorimoschema operculella)*; sweet potato weevil *(Cylas formicarius)*

The larvae feed on the grains and stored products such as, flour, oil cakes etc. (e.g.) Rice meal moth *(Corcyra cephalonica)*

The larvae and adults bore into the grains and feed on the inner contents (e.g.) Pulse beetle *(Bruchus chinensis)*; Rice weevil *(Sitophilus oryzae)*; Red flour beetle *(Tribolium castaneum)*; Lesser grain beetle *(Rhizopertha dominica)*. Some insects contaminate food products and cause damage (e.g.) Cockroaches *(Periplaneta americana)*, House flies *(Musca domesnica)*

**Insects causing indirect damage**

**i) Causing injury while egg laying.** Some insects while laying eggs cause injury to various plant parts by inserting the eggs into the tissues. The adult females make small cavities in the plant parts and lay eggs inside the cavities (e.g.) The mango nut weevil *(Sternochetus mangiferae)* makes small cavities in young mango fruits and lays eggs inside the cavities.

Many insects, especially the hoppers insert the eggs into the leaf tissues by making punctures with their ovipositors (e.g.) Mango hoppers *(Amritodes atkinsoni), Idioscopus niveosparsus, I. clypealis)*; Rice brown plant hopper *(Nilaparvata lugens)*; Rice green leaf hopper *(Nephotettix virescens, N. nigropictus)*

The adult female splits the branches of the host tree and lays eggs inside the injured portion (e.g.) Cicadas *(Platypleura machinoni)*

ii) **Using plant parts to build nests.** Some insects use plant parts to build nests in which they remain concealed and feed on the plant tissues.

A few insects cut pieces of leaves from the plants and use them to build nests for raising their young ones (e.g.) The leaf cutter bee *(Megachile anthracina)* cuts semi-circular pieces of leaves from rose or bean to build nests for their young ones.

Some insects web several leaves together to make box-like nests and live inside in colonies (e.g.) The red ants *(Oecophylla smaragdina)* commonly found in mango, cashew, guava and many other trees build such nests.

Some insects cut leaves and make cases out of them, remain inside the cases and feed on the leaf tissues (e.g.) Rice case worm *(Nymphula depunctalis)* cuts leaves of rice plants, makes tubular cases, remains securely inside, scrapes and feeds on the leaf tissues.

iii) **Other harmful aspects.** Severe infestation of some of the insect pests cause serious problems during the harvest operations.

a) **Problems caused during harvest.** Harvesting of produce from crops such as, beans, lab lab, soybean etc. severely affected by the Aphid *(Aphis craccivora)* and cabbage severely affected by the Cabbage aphid *(Lipaphis erysimi)* is very difficult.

In cotton crop, severely infested by Spotted boll worms *(Earias insulana* and *E. vitella)* as well as the Pink boll worm *(Platyedra gossypiella)*, collection of cotton lint from the bolls becomes very difficult.

Harvest of produce from fruit trees, such as mango, cashew, guava etc. infested by Red ants *(Oecophylla smaragdina)* poses serious problems.

b) **Deterioration of quality of the produce.** Infestation of plants or certain parts of the plants or the produce may lead to deterioration of the quality of the produce and consequently become unmarketable or fetch very low price in the market .

Infestation of some plants by certain pests results in heavy loss in yield and markedly affects the quality of the produce (e.g.) Cardamom plants infested by the thrips *(Sciothrips cardamomi)* causes heavy loss in yield and very badly affects the quality of the produce.

Borer infested sugarbeet, potato, egg plant, lady's finger, tomato, cabbage, cauliflower and such other vegetables, internode borer infested sugarcane, aphid infested beans, lab lab, cabbage etc., pest infested stored grains etc. are difficult to market and fetch very low prices.

c) **Retardation of growth of plants.** Most of the aphids and leaf hoppers secrete honey dew on the leaves and other parts of the infested plant. Sooty mould, a fungal disease develops on the honey dew and covers large areas of leaves and other plant parts. This dark fungal growth adversely affects the photosynthetic activity of the leaves resulting in retardation of growth of the plants.

   Infestation by large number of hoppers results in the development of sooty mould (e.g.) Mango hoppers *(Amritodes atkinsoni, Idioscopus niveosparsus* and *I. clypealis)*

d) **Attacking man and his livestock** Several insects such as, mosquitoes, bed bugs and lice attack man, suck and feed on the blood.

   Blood sucking flies, ticks etc. suck and feed on the blood of cattle.

   Fleas, bird lice etc. suck and feed on the blood of poultry birds.

e) **Transmitting diseases in man and live stock.** Many serious diseases such as, malaria, filariasis, typhoid, dysentery, leprosy, tuberculosis etc. are caused by insect transmitted pathogens. House flies, cockroaches etc. are also agents, which are responsible for spreading many diseases. Similarly insects transmit many diseases in livestock also.

f) **Transmitting diseases in plants.** Many insects serve as transmitters and vectors of several fungal, bacterial, viral and mycoplasma diseases

   A few fungal diseases are transmitted mechanically by some insects (e.g.) Bud rot, a fungal disease of coconut, arecanut and other palms is transmitted by the rhinoceros beetle *(Oryctes rhinoceros)*

   Sugary disease or 'ergot' disease of pearl millet caused by *Sphacelia microcephala* is mechanically transmitted by many insects that visit the flowers attracted by the sugary secretion, which comes out of the infected flowers.

   A few bacterial diseases are also transmitted mechanically by some insects. Bacterial wilt of cucurbits is mainly transmitted by the Red pumpkin beetle *(Raphidopalpa foevicollis)*

Most of the virus and mycoplasma diseases are transmitted only by insect vectors such as, aphids, white flies, hoppers, mites and thrips in a persistent or non-persistent manner (e.g.) Rice tungro virus disease is transmitted by the Green leaf hoppers *(Nephotettix virescens, N. malayanus, N. parvus* and *Recilia dorsalis)*; Red gram sterility mosaic virus disease is transmitted by Eriophyid mite *(Aceria cajani)*; Soybean mosaic virus disease is transmitted by the aphids *(Aphis craccivora, A. gossypii, A. fabae, Myzus persicae* etc.*)*; Bunchy top virus disease of Banana and Cardamom marble mosaic or 'Katte' disease are transmitted by the Aphid vector - *Pentalonia nigronervosa*; Papaya mosaic, Cucumber mosaic and Chilli mosaic diseases are transmitted by the aphid vector - *Aphis gossypii*; Vein clearing virus disease of Lady's finger, Tobacco leaf curl, Tomato leaf curl, Papaya leaf curl etc. are transmitted by the white fly - *Bemisia tabaci* ; Tomato spotted wilt virus disease is transmitted by the Thrips - *Thrips tabaci* ; Rice yellow dwarf mycoplasma disease is transmitted by the Green leaf hopper - *Nephotettix virescens*; Sugarcane mosaic virus disease is transmitted by the Corn aphid - *Rhopalosiphum maidis.*

g) **Transporting pests from one place to another.** Ants, some beetles and a few other insects, though do not cause any direct damage to the plants, help in transporting some pests from the infested plants to other plants mechanically. Mealy bugs, aphids and such other small insect pests, which are normally not very mobile are transported from one plant to another by sticking on to the legs or bodies of insect pests. Some species of ants afford protection to aphids in return for their honey dew secretion and carry on a symbiotic existence.

## Biotic or Natural balance

An **'ecosystem'** is a self-sufficient and self-regulated habitat, where the biotic and abiotic components of the environment interact together for the exchange of energy in a continuing cycle. The structural components of an ecosystem consist of **'producers'**, **'consumers'**, **'decomposers'** and **'abiotic factors'**. The producers are autotrophic organisms that utilize sunlight as energy and convert atmospheric carbon dioxide into carbohydrates through the process of photosynthesis. The consumers viz., insects, animals etc. derive their food directly or indirectly from the producers. The decomposers such as, fungi, bacteria etc. break down complex compounds of dead producers and consumers and convert them into simpler substances to be used again by producers. The abiotic components consist of non-living materials such as, water, minerals, atmospheric gases, salts etc. Thus, in an ecosystem there is a perfect equilibrium among the living and non-living components and the living organisms live in harmony in a state of compromise. This self-determined natural course is known as the **'biotic balance'**. This equilibrium usually prevents the multiplication of any one particular species of insect beyond a certain level, which may upset the biotic balance. Such a regulative process in an undisturbed environment neither allow decline of any plant or animal population to extinction nor allow the population to increase beyond a particular limit. The population levels of insects may fluctuate from year to year, season to season and from place to place, but the fluctuations are always restricted to within a particular limit and as such the biotic balance is not upset. However, during certain natural calamities such as, floods, cyclones, storms, volcanoes, severe drought etc., the biotic balance may be upset leading to flare up of some of the insects

The chief factors responsible for maintaining the biotic balance are, (i) Food, (ii) Climate (iii) Natural enemies and diseases.

Many forms of life, especially the lower forms often die of starvation if food is not available. If they happen to depend on particular plants for their food, the absence of such food arrests their growth and multiplication. Similarly extremes of climatic conditions also adversely affect the perpetuation of various animals and plants. The presence of various carnivores, which depend for their food on other forms of life are responsible for checking the multiplication of various smaller organisms. Besides such natural enemies, various diseases caused by fungal, bacterial, viral and other parasitic organisms also play a vital role in restricting the growth and multiplication of several organisms. When one or more of these checking factors are not prevented from playing their part unabated, then the biotic balance gets upset leading to flare up of some forms.

**Pest outbreak**

In certain seasons or in some months, the population of some insect pests increases suddenly to a large extent, over large areas and may cause extensive damage to the crops infested. Such incidences may happen because of certain changes in the weather conditions, especially sudden changes in the weather parameters such as, temperature, humidity rainfall etc. These sudden changes may sometimes become very favorable for the appearance and rapid multiplication of the pest and thus, the pest may become epidemic within a short time.

Besides such changes in the environmental conditions, availability of food, availability of growing space, lack of natural enemies etc. may also lead to pest outbreaks.

## Factors affecting pest outbreak

In nature, there is a perfect harmony between the environment and all organisms living in it, whether plant or animal. This is generally known as the **'balance of life in nature'**. The balance of life is usually maintained in a stable state except for small variations and as such, the population of each species tends to remain constant over long periods. However, when one or more of the important environmental factors are disturbed to a considerable extent, the normal equilibrium in nature is collapsed, as a result there may be abnormal increase in the population of some forms of insects, which become pests. One of the most important agency that causes such upsets in the balance of life is human beings. The advances of civilization has led man to conquer and control nature in various ways, as a result numerous changes have been brought about in the life conditions of almost all living organisms, especially the lower forms such as, insects. Such changes have resulted in occasional multiplication of certain forms of life, which ultimately results in pest out break. Some of the most important factors, which trigger to upset the balance of life in nature are :

i) **Sudden changes in weather conditions.** Certain occurrences in nature may cause sudden changes in the environmental conditions, which may be quite favorable for the multiplication of some forms of insects leading to flare up of these insect population to epidemic level within a short span of time.

ii) **Destruction of forests.** Forests play a vital part in maintaining the climatic conditions of a place at a constant level. When forests are destroyed by man for various purposes the environment is affected. Consequently the quantum of rainfall is reduced and the temperature increases. Such changed atmosphere may be favorable for some pests, which may increase in number at an alarming rate.

iii) **Destruction of forests for cultivation of crops.** The most important requirement for the existence of life is food. When forests are destroyed for cultivation of crops, several insects originally living on forest vegetation, transfer their activities to cultivated crops and become serious pests in due course.

iv) **Destruction of natural enemies.** The reckless destruction of insectivorous birds, reptiles and mammals, as well as extermination of entomophagous insects such as, parasitoids and predators knowingly or unknowingly by indiscriminate use of plant protection chemicals interfere with the functions of such organisms, which act as natural checks of several pests.

v) **Adoption of intensive cultural practices.** Several measures to increase agricultural production have led to outbreak of many insect pests. Many high yielding varieties of crops are being introduced periodically to increase crop yield. Many of these varieties are not resistant to important pests which infest them. Cultivation of such susceptible varieties lead to flare up of several insect pests.

Similarly many high nitrogen tolerant varieties are being introduced to increase crop yield. Application of large quantities of nitrogenous fertilizers makes the plants to grow luxuriantly with rich foliage, inviting pests infesting the crop, which may multiply very rapidly because of the availability of plenty of food.

Reduced spacing and close planting of crops provide a micro-climate within the crop canopy, which may be conducive for the rapid multiplication of some of the pests.

Growing of single crops in the same locality in extensive areas helps pests feeding on such crops to increase and multiply rapidly. Cultivation of a single crop or a single variety in a locality continuously for many seasons helps pests infesting the crop to establish permanently in that locality and leads to serious pest outbreak.

vi) **Introduction of crops in a new locality.** When a crop is freshly introduced in a locality where it has not been cultivated previously, the pests, which had infested the crops cultivated in that locality previously start attacking the new crop and ultimately the new crop becomes the primary host of the pests and multiply rapidly.

vii) **Introduction of pests in new localities.** Pests, which had not been found in a particular locality, may be introduced afresh knowingly or unknowingly through seeds, seed materials or soil. When the new environment is found to be favorable for the introduced pests, they may multiply rapidly and become serious pests.

viii) **Introduction of new pests from foreign countries.** Pests from one country may be transported to other countries by road, rail, seas or air through seeds, seed materials, plants, plant parts etc. by passengers knowingly or unknowingly. The pests thus newly introduced may become serious pests in the new environment and in the absence of their natural enemies in the country. The San Jose scale, Mediterranean fruit fly, cotton boll weevil, sugar beet weevil, sugarcane beetle, potato golden nematode etc. are pests, which have been introduced to India from foreign countries.

ix) **Occurrence of pests resistant to insecticides.** Repeated application of some insecticides and application of insecticides at incorrect dosages may result in the development of resistance in pests to such insecticides. The resistance may be genetically controlled or may develop as a result of mutation of some genes in the pests. This type of resistance may be transmitted to subsequent generations of the pests. The pests, which have developed such resistance to insecticides cannot be destroyed even at much higher concentrations of the insecticides and may multiply very rapidly unabated.

## Pest monitoring

Many insect, as well as non - insect pests attack cultivated crops. In addition to the regular and permanent pests, sometimes the seasonal pests may also infest the crops on a large scale leading to severe loss in yield. Knowing the details of the pests infesting the crops such as, nature of the pests, their life history, nature of damage caused by them, the vulnerable growth stage at which they can be controlled effectively and suitable control measures, which will go a long way in adopting appropriate measures to control them is known as **'pest monitoring'**. Collection of necessary informations for this from cultivated fields is known as **'pest surveillance'**.

## Forecasting pest outbreak

Based on certain changes in the weather conditions and the initial appearance of the pests, it may be possible to forecast flare ups of certain insect pests on a large scale. Such forecasts may give a chance to mobilize all resources, so as to take up appropriate control measures before the occurrence of the pest on a large scale and cause extensive damage to crops. This type of forecasting is possible based on different observations :

i) **Forecasting based on the initial appearance of the pest**. Initial appearance of the pests may be ascertained by the use of light traps, pheromone traps, poison baits, sticky traps etc. and based on the number of insects collected by such gadgets, the chances of large scale occurrence of the pests may be predicted. Such forecasting is possible in the case of pests such as, rice stem borer, rice brown plant hopper, groundnut red hairy caterpillar etc.

ii) **Forecasting based on pest surveillance in cultivated fields**. Informations gathered regularly from fixed plot surveys, as well as from roving surveys of the pest surveillance programs in cultivated fields provide valuable information on the appearance and build-up of pests. Based on the increasing trends in the pest population, the possibility of large scale occurrence of certain pests may be forecasted.

Such forecasting is possible in the case of pests such as, rice green leaf hoppers, cotton boll worms, cotton white fly etc.

iii) **Forecasting based on weather parameters**. There is close correlation between the weather conditions and pest outbreaks. In the pest surveillance programs, details of weather parameters such as, maximum and minimum temperature, relative humidity, amount of rainfall, number of rainy days, wind velocity etc. are collected regularly. Certain changes in the weather conditions may be favorable or unfavorable for the multiplication and build-up of some of the pests. The weather parameters and the incidence of pests are correlated statistically and the possibility of large scale occurrence of the pests with the changes in weather conditions may be predicted. Steady increase in temperature is unfavorable for the multiplication of many insect pests. Low temperature and high atmospheric humidity continuously for long spells is found to be favorable for large scale occurrence of rice green leaf hopper. Similarly low temperature and high humidity is favorable for the multiplication of aphids. Heavy rainfall brings down the population of thrips to a considerable extent as these insects are washed down due to heavy rainfall.

iv) **Forecast based on the locality**. All insect pests do not appear uniformly in all locations. Some insect pests continue to appear regularly in certain localities in all the seasons and in all the years and may cause serious damage to some crops. Based on this fact, the possibility of some of the serious crop pests occurring in certain localities on a large scale whenever the hosts are cultivated may be forecasted. In South Arcot district, large scale leaf miner infestation is found in the rainfed groundnut crop every year regularly, In Thanjavur district, rice crop may be severely affected by stem borer infestation during the Thaladi season. In some specific localities, rice earhead bugs appear regularly in all the seasons. Every year, after the receipt of a few good summer showers, groundnut red hairy caterpillars appear in Coimbatore and Pollachi districts.

Whenever possibilities of such large scale occurrence of serious crop pests are predicted, the information is given wide publicity through various media such as, news papers, radio, television etc., so that precautionary measures may be taken well in advance to combat the pest menace as they appear.

There are two kinds of forecasting viz., **'Short term forecasting'** and **'Long term forecasting'**. Short term forecasting covers only a particular season or one or two consecutive seasons. This is possible by *in situ* studies on the pest population by employing insect trapping methods or other such sampling techniques within the crop. Such predictions can also be made based on the rate of emergence of the pest under laboratory conditions. Long term forecasting covers wide area and is mainly based on the possible effects of weather parameters on the occurrence of pests in epidemic proportions.

### Forewarning of pest occurrence

When a serious crop pest appears in an epidemic form in a village, district or state, there is every possibility that the same pest may appear in the adjoining villages, districts or states respectively. In such cases the information about the possibility of the pest moving to the adjoining villages, districts or states is conveyed through various media such as, news papers, radio, television etc. as a warning, so that effective precautionary measures may be taken up to prevent large scale occurrence of the pest. This is known as **'forewarning'**. Such forewarnings are given when serious crop pests such as, rice brown plant hopper, rice green leaf hopper, which is also the vector of the virus causing rice tungro disease, groundnut red hairy caterpillar, cut worms, locusts etc., which occur suddenly in some places in an epidemic form.

### Assessment of insect population

To study the extent of damage that may be caused by insect pests and to adopt control measures at the appropriate time, it is quite necessary to assess the population density of the pest. In pest population assessment, it may not be possible to count all the insects present in the crop in the field. So, the assessment has to be made from representative random samples. Assessment is mostly done at the injury causing stage of the pest viz., larvae in the case of Lepidopterous pests, nymphs in the case of bugs, adults in the case of beetles, both nymphs and adults in the case of earhead bugs, grasshoppers etc. Different methods are being followed for the assessment of pest population.

i) **Net sweepings**. This method is adopted to assess the population of insect pests such as, adult beetles, nymphs and adults of bugs, grasshoppers etc. An insect net with a loop of 43 cm. diameter and 1.0 m. long handle is used for this purpose. Usually 10 sweeps are sufficient to assess the general population of any species of insect and is expressed in terms of number of insects per 100 sweeps. A sweep consists of half a circle or 180° with the net held in a vertical plane.

ii) **Sudden trapping**. In this method, the insects in an unit area are suddenly trapped by means of suitable traps without causing any disturbance and the trapped insects are counted. This method is followed in the case of certain grasshoppers, sorghum earhead bugs, insects belonging to Collembola etc.

iii) **Sight counting**. The population of pests in a measured area is counted directly and expressed as number of insects per unit area. This method is followed in the case of caterpillars of moths, butterflies etc., which are external feeders and can be easily seen with naked eye.

iv) **Light trap catch**. This method is useful to assess the seasonal abundance of many species of insects, which are attracted to light. The brood emergence of groundnut red hairy caterpillar, rice yellow stem borer etc. are monitored by this method. Large scale occurrence of pests such as, rice green leaf hopper is also monitored by this method.

v) **Sticky trap** or **Adhesive trap catch**. Sticky traps are set up and flying insects are trapped on the sticky material. Sticky traps are commonly used in migratory studies of insects such as, aphids, white flies etc. Yellow sticky traps are widely used for trapping adults of Aleyrodids such as, *Bemisia tabaci* and *Trialeurodes vaporariorum.*

vi) **Bait trapping**. Some raw plant materials and some synthetic chemical formulations serve as attractants. The sexual odor or food odor produced by such substances lures the insects to the source of such substances. These substances are used as baits in special types of traps to attract the insects, which can be trapped. Methyl eugenol is a chemical sex attractant for the mango fruit fly - *Bactrocera (Dacus) dorsalis* and *B. zonatus.*

vii) **Water trap**. A floating pan trap is used for trapping insects found on water surface.

vii) **Suction trap**. Sucking in air into a trap with a sucking device operated either by hand or by means of a motor, traps flying insects. Hourly catches of flying insects are made possible by means of a timing machine.

ix) **Narcotized collection**. This method is followed to assess the population of quick moving insects. The insects are killed or immobilized by spraying quick acting, non - persistent chemicals such as, dichlorvos, pyrethrum etc. or anaesthetized by the use of chemicals such as, chloroform, ether etc. The insects, which fall down are collected on a sheet of paper spread below the plant and counted. The population of mirid bugs on fruit trees is assessed by this method.

x) **Soil area** or **volume count**. Soil-inhabiting insects such as, root grubs can be assessed by this method. Soil samples are collected from unit areas up to a fixed depth and the insects present in the samples are counted and the insect population is expressed in terms of number of insects in unit area or unit volume of soil.

- **xi) Emergence cages**. Emergence cages covering unit areas are set up and the number of adult insects emerging from unit areas is assessed.
- **xii) Extent of damage**. The extent of damage caused by an insect pest is sometimes taken as an index for the assessment of the pest population.
- **xiii) Marking and recapturing**. A large number of insects are caught by means of insect traps and are marked with paint or tagged. Then they are released in the field and allowed to mix with the general population. The insects are later trapped at different distances from the point of release. The tagged, as well as the other insects collected help in assessing the population, extent of multiplication and activity of the pest. Of late, radioactive tracers are used to trace insects. This type of study provides valuable information about the rate of dispersal, flight range and migration of some of the insect pests.

## Sampling

The important criteria in sampling are the distribution of the pest in the crop, nature of the sample, the method of sampling, the size and number of samples, population assessment and analysis of the sample counts. Insect counts can be taken from the air over a crop canopy, on the infested plants or from the soil. The nature and number of samples to be taken depends largely on the insect, growth stage of the insect and its distribution in the host. It is necessary to fix a minimum number of samples. However, it is always better to have a large number of samples in order to have a more precise estimation of the pest population.

In assessing pest population, the sample unit, which are infested plants, have to be selected at random and this should be done in an unbiased manner. Intentional selection of samples with more number of pest or less number of pest should be avoided. The population of the insect in the crop samples is counted and recorded. This method is widely followed in the case of insect pests such as, rice stem borer, sorghum stem borer, gall midges, aphids, thrips, boll worms etc. However, the techniques for assessment of population differ depending upon the crop and the pest species. In the case of cotton, for assessing the incidence of boll worm in a field, a specified number of plants in each row along a diagonal from the corners of the field are examined. The percentage of boll worm attack is assessed based on the total number of bolls and the number of infested bolls on these plants. Incidence of aphid and leaf hopper is assessed based on the population of the insects present on the leaves selected from the terminal, middle and basal portions of the plants.

In the case of densely populated insect pests, lesser number of samples of smaller size may be sufficient, while in the case of thinly populated insect pests, the number of samples and the size of the samples have to be increased.

## Estimation of damage caused by insect pests

In general the damage caused to crops depend to a large extent on the pest population. The higher the pest population, the more is the damage caused by them. However, in the case of insect vectors of virus and mycoplasma diseases, even a very small population is quite sufficient to cause heavy crop losses. The damage caused by insect pests is ultimately reflected in loss in yield. The loss may be estimated quantitatively in terms of yield in kg / acre in the case of many grain crops or quantitatively in terms of infested and spoilt fruits and vegetables. The methods employed to interpret the relationship between pest population, crop damage and consequent yield loss and ultimately the economic loss vary very much depending upon the pest and the crop affected.

i) **Estimation from visual observations**. Rough estimate of yield losses due to infestation by many pests can be gauged by general visual observation of the infested crop. The annual loss due to pest infestation in India is roughly estimated at 10 per cent.

ii) **Estimation based on field surveys**. By adopting standardized and widespread surveys, the loss incurred due to a pest species can be worked out. The data thus collected from surveys on a number of representative areas from different parts of the country for several consecutive years and properly scrutinized may give a fairly reliable estimate of damage caused and the resultant yield loss due to a wide range of pests. The data collected will also be useful in long term forecasting of occurrence of some of the serious pests in epidemic proportion, which occur in cycles when identical conditions recur.

iii) **Estimation from experimental plots**. This is a more accurate method to evaluate the extent of pest damage and the resultant yield loss. In this method, the pest under study is controlled to the maximum possible extent by different control practices including application of appropriate insecticides. The yield obtained from such crop is compared with the yield obtained from a crop subjected to natural infestation. To have an accurate estimate of damage caused by the pest and the consequent yield loss, a large number of experiments under diverse conditions and localities have to be conducted. This method is also faced with many constraints:

   **a)** Application of an insecticide to keep the crop free from a particular pest may also kill other pests that may attack the crop. Under such conditions the damage caused by the particular pest alone and the consequent loss in yield cannot be ascertained correctly. To avoid this, a selective insecticide, which controls the particular pest has to be used.

**b)** By the application of a selective insecticide, if 100% kill of the pest is not realized and if parasitoids and predators attacking the pest are also killed, then the population of the pest increases.

**c)** When a target pest is controlled, that may sometimes lead to flare up of other pests attacking the crop and may cause more damage than under normal circumstances and thus may vitiate the results.

**d)** For conducting this type of experiments, it is quite necessary that the experimental plots should be of optimum size and are set far apart.

**e)** In all such experiments seeds from the same lot from a selected variety should be used.

**f)** The insecticide selected should not be phytotoxic to the crop and it should not have any growth promoting effect on the crop.

**iv)** **Estimation from cage experiments**. A known number of the pest species to be studied is allowed on a single plant or an unit crop area and caged so as to exclude infestation by the same pest or pests from the outside. This method may give a relatively correct estimation of damage caused and the yield loss. The disadvantage in this method is that the plant or crop growth may be affected to a certain extent due to caged condition.

**v)** **Direct estimation from field crops**. In a field infested by a particular pest not all the plants or plants in unit areas are infested to the same level. Plant or plants in unit areas of the same field are infested to varying degrees. While some are severely infested, some are moderately infested and some may even be free from infestation. The yield obtained from individual plants or plants in unit areas exhibiting different degrees of pest infestation, as well as from uninfested plants are recorded separately. Such data collected from different fields are subjected to statistical scrutiny and a correlation equation is worked out between the degree of pest infestation and yield. The yield loss under different degrees of pest infestation can be obtained from this equation.

## Economic threshold level

When the population of an insect pest and the damage caused by it do not rise beyond a certain level, there may not be economic loss. Hence economic threshold level is the infestation level beyond which level there will be economic loss. Plant protection measures have to be taken up as soon as the pest infestation reaches the economic threshold level, so that economic loss may be avoided.

The economic threshold level of some of the most important crop pests are given in **Appendix - 2**

## Economic injury level

Economic injury level is the infestation level at which the damage caused by the pest will result in economic loss. So, economic injury level is always higher than the economic threshold level. By taking up plant protection measures at the economic threshold level, the infestation may not reach the economic injury level and thus economic loss can be avoided.

## Classification of insect pests

Insect pests are classified in different ways depending upon the extent of damage, nature of occurrence, biological nature, severity of attack etc.

### 1. Classification based on the extent of damage

i) **Major** or **Key pests.** Infestation by such pests may cause severe damage to the crop and the yield loss may go up to 10 percent or more. Infestation by these pests may result in total destruction of the crop sometimes (e.g.) Rice stem borer; Rice brown plant hopper; Egg plant shoot and fruit borer.

ii) **Minor pests.** Infestation by such pests may not cause severe damage to the crop and yield loss may be from 5.0 - 10.0 per cent (e.g.) Red cotton bug; Rice white backed hopper; Groundnut thrips; Saw toothed grain beetle.

### 2. Classification based on nature of occurrence

i) **Regular pests.** Some insect pests infest a particular crop whenever that crop is cultivated, irrespective of the season or months. These pests always have a close link with the particular crop and infest the crop positively any time before harvest of the crop. Generally the major pests come under this category (e.g.) Rice stem borer; Groundnut leaf miner; Egg plant shoot and fruit borer.

ii) **Seasonal pests.** These pests infest certain crops only during certain months or in some particular season. The suitability of the weather factors prevailing during the months or in that particular season is responsible for the appearance and multiplication of the pest (e.g.) Castor semilooper appears only during the months of August - January; Groundnut red hairy caterpillar appears in Coimbatore and Pollachi

Districts of Tamil Nadu only during the months of June - July after heavy summer showers.

iii) **Occasional pests.** Unlike regular and seasonal pests, these pests infest certain crops occasionally. There is no definite link between the pests and the crops they infest (e.g.) Rice case worm; Castor slug caterpillar.

iv) **Sporadic pests.** These pests sometimes infest certain crops within some limited areas (e.g.) Rice ear head bug.

v) **Persistent pests.** These pests remain in the crop all through the year permanently and continue to cause damage. These pests are found mostly in perennial crops (e.g.) Mango scale insect; Cardamom thrips.

vi) **Potential pests.** These pests do not cause much damage usually. But sometimes, when the environmental conditions are highly suitable for the pests, they may assume serious proportions and become a major pest (e.g.) Rice leaf folder.

vii) **Migratory pests.** These are pests, which under certain conditions migrate in very large numbers in swarms, feed and destroy all the vegetations on their way (e.g.) Desert locust

### 3. Classification based on the biological nature

i) **'r'- pests**. The pests of this category are generally small to medium sized insects. They are capable of laying large number of eggs, complete their life cycle within a short period, thus multiply and spread very fast and can cause extensive damage to crops they infest. Grasshoppers, most of the Lepidopterous insects, aphids etc., are included in this category (e.g.) American cotton boll worm - the female moth lays 300 - 1000 eggs and the whole life cycle is completed in 30 - 60 days and several overlapping generations may appear in a year; The aphids, which infest cruciferous plants, the female insect is capable of viviparous reproduction and the nymphs become adults within a few days and start reproduction and can multiply very rapidly. There may be several generations in a season.

ii) **'k' pests**. The pests of this category are usually large-sized. They lay lesser number of eggs and it may take several months to complete the life cycle of the pest. But they are very hardy insects and can tolerate and withstand adverse environmental conditions (e.g.) Coconut rhinoceros beetle - The female beetle lays only 100 - 150 eggs and the life cycle is completed in 220 - 290 days. Only one generation may appear in a year; Mango stem borer - the female beetle lays only less number of eggs and the life cycle lasts for 8 - 9 months. Two generations may appear in one year; Coconut red palm weevil - the female lays 200 - 250 eggs and the life cycle is completed in 75 - 135 days.

iii) **Intermediate pests**. These pests are mostly intermediate between the 'r' pests and 'k' pests in all respects (e.g.) Castor hairy caterpillar - The female moth lays 120 - 220 eggs. There may be 7 - 8 generations in one year. The life cycle is completed in 45 - 60 days.

**4. Classification based on severity of attack**

i) **Epidemic pests.** The pests of this category appear suddenly in very large numbers in certain regions during a particular season and cause extensive damage within a very short period of time (e.g.) Rice brown plant hopper; Cut worms.

ii) **Endemic pests.** These pests are always found to infest certain crops, whenever they are grown in some places to a lesser or greater extent, depending upon the environmental conditions (e.g.) Rice leaf folder, Groundnut leaf folder.

# 2

# Insect Pest Control

**'Insect pest control'** includes all measures adopted to make the life of insects hard and ultimately to destroy them or to prevent their multiplication and spread. Such measures can be taken up before the appearance of the pests or after occurrence of the pests. To achieve this purpose , it is quite necessary to know the factors, which favor the occurrence of the pests and acquire enough knowledge about the nature and life cycle of the pests, the vulnerable growth stage at which they can be effectively destroyed and appropriate measures of control to destroy them Control measures may be either preventive or direct.

## Need for pest control

Control measures for the control of pests have to be taken up as soon as the infestation reaches the economic threshold level. Failure to control the pests at this stage will result in further increase in the pest population and consequent loss in yield and economic loss.

The need for pest control is of two kinds :

i) **For yield maximization.** The only objective of pest control in this method is to maximize the yield. The cost incurred for pest control is most often more than the income accrued as a result of pest control. The method is adopted only with the object of increasing the yield as much as possible so that the country and the people may be benefited. In this method, pest control operations are taken up before the actual occurrence of the pest or before the pest infestation reaches the economic threshold level. Such treatments taken up to control the pests before their occurrence are known as **'prophylactic treatments'**

ii) **On the basis of cost / benefit ratio.** The main aim of this method is to obtain higher yields, at the same time to get a margin of profit as a result of pest control. In this method, the cost incurred towards pest control is either equal or lesser than the monetary benefit obtained by way of yield increase due to pest control. Here pest control operations are taken up only after the pest infestation and the resultant damage reaches the economic threshold level. This is known as **'need - based pest control'**.

**'Insect pest control'** is broadly classified under two different heads viz., **'Natural control'** and **'Applied control'**.

## I. Natural control

1. **Climatic conditions** - Temperature, rainfall, sunshine etc.
2. **Physical** or **topographic factors** - Mountains, oceans, rivers, lakes, forests, deserts etc.
3. **Natural enemies** - Parasitoids, predators including insects, birds, mammals, reptiles etc
4. **Disease causing organisms** or **parasites** - Entomophagous fungi, bacteria, viruses and other microorganisms.

## II. Applied control

### 1. Prophylactic measures

i) **Clean cultivation** - Elimination of weeds and other collateral and alternate hosts, destruction of pest infested plant parts fallen on the ground, pruning, manuring, thinning, irrigation etc.

ii) **Cultural practices** - Deep ploughing, summer ploughing, crop rotation and other cultural practices

iii) **Cultivation of resistant varieties** - Cultivation of pest resistant or pest tolerant varieties.

iv) **Other methods** - Seed treatment, sun drying of seeds, tree banding etc.

### 2. Curative or Direct methods

i) **Cultural methods** - Removal and destruction of infested plant parts, trash mulching, earthing up, impounding water etc.

ii) **Physical methods** - Hand picking, bagging, netting etc.

iii) **Mechanical methods** - Poison bait traps, light traps, pheromone traps, heat treatment, radiation etc.

iv) **Use of natural agencies** - Sun drying, flooding etc.

v) **Legal control** - Government regulations for the control of important pests, pest quarantine laws etc.

vi) **Biological control** - Protection of natural enemies such as, parasitoids and predators, insectivorous birds, mammals, reptiles and such other predators, introduction of natural enemies of insects, use of parasitic microorganisms such as, fungi, bacteria, viruses etc.

vii) **Chemical control** - Use of insecticides viz., stomach poisons, contact poisons, fumigants, attractants, repellants etc.

viii) **Other modern methods** - Use of radiation, chemosterilants, antibiotics, sulfanilamides, hormones, genetic manipulation, antifeedants or feeding deterrents etc.

ix) **Integrated control measures** - All possible control measures taken up simultaneously

## I. Natural control

Several inter related forces of nature such as, climatic factors, topography, natural enemies etc. play vital roles in maintaining the population of all living organisms at a stable and balanced equilibrium. Such natural factors cannot be altered, manipulated or controlled by the activities of man. These natural control factors often act as checks on the insect population, their distribution and spread.

### 1. Climatic conditions

**'Climate'** is the most important natural factor, which keeps insect populations in check. These factors include temperature, rainfall, sunshine, humidity, wind velocity, atmospheric pressure etc., which individually or collectively affect every phase of the life of insects directly or indirectly. A warm, moist environment is favorable for the development of most insects. Extremes of temperature are destructive to many insects. Cold and wet winter weather and sharply fluctuating temperature changes destroy many insect species. Optimum temperature for the activities of most of the insects is 26°C and aestivation starts at 38°C.

Sufficient rains and the consequent softening of soil is highly favorable for the groundnut red hairy caterpillar, tobacco cutworms, black hairy caterpillar etc. to enter into the soil for pupation and for the adult moths to emerge out from the pupae inside the soil. On the other hand, heavy rains may destroy many insects such as, thrips, aphids etc. by washing them down to the soil. Stagnation of water may also lead to destruction of many insects dwelling in the soil. Cloudy and dark days may adversely affect the activities and spread of many insects. Sunshine also plays an important role in the life of most of the insects, as they are either nocturnal or diurnal in habit. Some insects fly about in bright sunshine only, whereas it is unfavorable for many other insects. Dispersal of several insects is favored by wind as they generally fly with the air currents, although some insects such as, mosquitoes fly against air currents. Strong winds besides causing hindrance to mating and egg laying in many insects, may also destroy many weak-flying insects.

## 2. Physical or Topographic factors

Physical factors such as, high mountain ranges, Oceans, lakes, rivers, deserts, forests etc. may act as natural barriers and prevent the spread of insects across them. Further, such agencies may influence the climatic conditions to a large extent and as such, control the development and spread of many insects. The physical factors may affect the soil conditions also. Physical and chemical nature of soils may affect the insects inhabiting the soil. The ragi root grub *(Holotrichia consanguinea)* is found more often in loamy soils, whereas cumbu root grub *(Anthrodeis* sp.*)* is more prevalent in sandy soils. The soil is the deciding factor for the type of crops grown in it and in turn the pests inhabiting them. The quality of water also affects insects to a considerable extent . Stagnant water pools harbor mosquitoes, some flies, beetles etc. to a large extent than free flowing water

## 3. Natural enemies

Each and every insect species has its own natural enemies. Parasitoids, predators, spiders, birds, mammals, reptiles, fishes etc. are the most common natural enemies of insects. They serve to keep the population of insects at a balanced level. When there is an increase in the insect population, their natural enemies also increase in number and attack and destroy the insects, thus help to bring down the insect population . When the insect population goes down, the number of their natural enemies also gets reduced automatically, thereby a balance is maintained.

**'Parasitoids'** are insects, which attack different growth stages of other insects such as, eggs, larvae or pupae and destroy them ultimately. Similarly there are many insect predators such as, preying mantis, dragon flies, damsel flies, mole crickets, predatory bugs etc., which catch and devour other insects. Many birds are known to catch and devour insects such as, locusts, grasshoppers, worms, tree - boring larvae, scale insects, white ants etc. Wood peckers , king crows and a few other birds live exclusively on insect diet. Mynah, crow, fowl, duck, stork, crane, owl, fly catcher etc. also feed voraciously on terrestrial insects. Non-insects such as, frogs, toads, lizards, chameleons, bats, snakes, spiders etc. are also insect feeders. Some nematodes are also known to attack and destroy a few insect species. Fishes feed on aquatic insects and larvae. Some caterpillars are known to eat other caterpillars of their own species under some circumstances and bring down their own population. This is known as **'cannibalism'**. The green caterpillar *(Helicoverpa armigera)* and the sunnhemp pod borer *(Etiella zinckenella)* come under this category.

### 4. Disease causing organisms or Parasites

Besides parasitoids and predators, some parasites such as, entomophagous fungi, bacteria, viruses etc. attack several crop pests, cause diseases and ultimately destroy them.

## II. Applied control or Artificial control

Without the efforts of man, the natural forces help to maintain the insects in check and keep them at a balanced level. But too much of man's interference in nature tends to upset this balance, resulting in large scale occurrence of some of the insect pests, which attack and damage the crops raised by man to meet his requirements of food, clothing and shelter. Under these circumstances, man ventures to control these pests by applied methods. Unlike factors, which affect natural control, the artificial measures are apt to be controlled or manipulated by man to a large extent These measures may be taken up prior to the appearance of the pest, which is known as **'preventive'** or **'prophylactic methods'** of control and the measures taken up after the occurrence of the pest, is known as **'curative'** or **'direct methods'** of control.

### 1. Prophylactic methods of control

These methods are quite effective in controlling insect pests, which occur season after season or year after year regularly.

i) **Clean cultivation**. Several grasses and weeds serve as alternate or collateral hosts of many insect pests in the absence of their normal cultivated crop hosts. Regular and periodical removal of weeds from field bunds, irrigation channels and surrounding areas of fields help in reducing the pest population to a large extent. In the absence of rice crop, rice mealy bugs, rice green leafhoppers etc. continue their life in several other grass hosts. Similarly, rice earhead bug has many other collateral hosts.Infested leaves, flowers, fruits etc. fallen on the ground should be removed periodically and destroyed, as larvae and pupae may be present in them. Population of many borer pests, such as cotton boll worms, mango fruit borer, egg plant fruit borer, bud worms etc. may be reduced in this way.

   Dead grasses and leaves should be collected and burnt to kill insects, which may find shelter and protection under them. Cutworms generally hide in such places during the day time. Many insects hide under dense grasses, fallen leaves and other dead vegetation during the winter season.

   Coconut rhinoceros beetle larvae live inside such refuse heaps. Crop residues, stubbles etc. should be removed and destroyed, as they may harbor insect pests. Rice stem borer may continue to live in the stubble.

Cotton pests may continue their lives in plants left over in the field after completion of harvest of the lint. Termites may thrive on sugarcane stubble.

ii) **Cultural practices**. Some of the cultural practices adopted serve to prevent the occurrence of some of the pests and thus are prophylactic in nature. However, most of the cultural methods are both prophylactic and curative and cannot be differentiated.

One of the most important method is selection of pest-free seeds and seed materials. Several seed-borne pests such as, cotton boll worms sugarcane scale insects and mealy bugs, banana stem weevil, sugarbeet weevil etc. may be avoided by this method.

Another method is raising of early maturing varieties of crops, which may escape pest infestation. Early maturing cotton crop is not subjected to attack by pink boll worms.

iii) **Cultivation of pest resistant varieties**. Growing resistant varieties is by far the best method of pest control. However, most of the resistant varieties do not possess desirable agricultural qualities. Further, the resistance is not always stable under different conditions of soil and climate. Resistance may be due to morphological, chemical or physiological conditions of the host plants. Some varieties of crops are not attacked by some of the pests or the pests do not cause any appreciable damage to the varieties. This is known as **'resistance'**. Some varieties of crops are able to withstand pest infestation and still able to grow and perform almost normally. This is known as **'tolerance'**. This may be due to some physical or physiological characters in the hosts. Some varieties of crops possess certain toxic metabolites or they may not have the required nutrients for the pests. Pests may shun such varieties. This type of resistance is known as **'antibiosis'**. Some varieties of crops may not have the quality, color, odor or taste preferred by the pests or they may possess thick and hard epidermis, waxy coating, hard salty deposits or dense hairs. Pests usually avoid such varieties. This is known as **'non-preference'**. At any rate, in India, no pest problem has been successfully solved altogether by the use of resistant varieties.

iv) **Other prophylactic methods**

a) **Seed treatment**. Mixing sorghum seeds with carbofuran at 1.5 kg. of carbofuran 3 G per kg. of seed (5% ai.) or treating the seeds in a solution of imidacloprid 70 WG. @ 10 g. ai. / kg. of seed prior to sowing protects the seedlings from shoot fly infestation. Treating paddy seeds in hot water at 50° - 55°C for 15 minutes kills the nematodes present inside the seeds.

b) **Seedling treatment**. Dipping rice seedlings in a solution of chlorpyriphos 20 EC. at 2.0 ml. / litre of water (0.04% ai) prior to planting affords protection from stem borer, gall fly and green leaf hopper infestation for about one month.

c) **Tree banding**. Painting tree trunks with a mixture of kerosene oil and tar prevents white ant attack. This is usually adopted in the case of coconut trees. Washing the stem portions of coffee plants with carbaryl at 20 g. of carbaryl 50 WP per litre of water (1.0% ai.) affords protection against coffee white borer attack.

d) **Sticky bands**. Use of sticky bands around the trunks of trees prevents insects from crawling up the trees. This method is adopted against mango mealy bugs.

e) **Stirring manure pits**. Stirring manure pits periodically brings out coconut rhinoceros beetle larvae from the deeper layers, which are subsequently destroyed by direct exposure to sun's heat or eaten by insectivorous birds or other predators.

f) **Treating of harvested grains**. Harvested grains are thoroughly dried in sun and then stored to prevent infestation by stored grain pests

g) **Treating stagnant water**. Adding small quantities of kerosene oil or crude oil over stagnant pools of water destroys mosquito larvae and pupae.

h) **Ploughing**. Ploughing the stubble after harvest and burying them under the soil or burning destroys stem borers of rice, sorghum etc.

i) **Flooding**. Flooding the fields with water kills a number of insects, which live and breed in the soil.

j) **Adjusting sowing time**. Altering the sowing time by either advancing or delaying it is advocated in some cases, so that the vulnerable stage at which the crop is subjected to attack by the pest may not synchronize with the period when the pest becomes abundant.. Red gram crop grown in the month of May, two months before the onset of the South-West monsoon rains, escape pod fly infestation.

## Curative or Direct methods of control

These measures are taken up after the occurrence of the pest, either to destroy them or to prevent their multiplication and spread.

### i) Cultural methods

Some of the cultural methods, when adopted at the appropriate time either destroy a large number of pests or prevent their multiplication to a considerable extent or reduce their population to some extent. The cultural practices, which are adopted to control a particular pest, may not be effective in controlling other insect pests.

a) **Crop rotation**. Crop rotation is one of the most frequently recommended methods to check the multiplication of insect pests. If a crop is cultivated in the same field for years together in succession, it affords a very favorable atmosphere for the multiplication of insect pests associated with the crop, as they are able to get a continuous supply of food and their multiplication goes on uninterrupted. When sugarcane crop is cultivated in an area continuously, the pests attacking the crop go on increasing year after year. In the rice belt of Thanjavur District in Tamil Nadu, because rice is cultivated continuously, the pests infesting rice crop is found throughout the year in abundance and cause extensive damage. On the other hand, if a crop grown in a field for one year is followed by a different crop, the pests dependent on the first crop may not be able to live and multiply on the second crop and for want of their natural food and suitable environmental conditions may perish. Sometimes crop rotation may be the only alternative for controlling certain pests. However, crop rotation as a means of controlling pests has certain limitations. The second crop succeeding the first crop in the rotation schedule should not be susceptible to the pests, which infests the first crop. Cotton crop should not be followed by lady's finger crop, because most of the pests infesting cotton infest lady's finger crop also.

   Crop rotation is successful only when the pest to be controlled has a limited host range. Controlling polyphagous insect pests by crop rotation is a non - practicable proposition. Further, the pests to be controlled by crop rotation should be non-migratory in nature and should have a comparatively long life cycle.

b) **Trap cropping**. To control insect pests, which infest more than one crop, trap cropping may be adopted. Around the primary or main crop or in-between the primary crop, as row crop or near about the field where the primary crop is grown, some other crop of lesser importance (trap crop), which is also a favorite food of the pest may be raised. The pest, which is attracted by the trap crop may be destroyed by the application of suitable insecticides or by other methods and thus, the primary crop may be saved from severe pest infestation. Castor or red gram may be raised as trap crop along with groundnut to control red hairy caterpillar and *Prodenia* caterpillar. The pests, which are attracted by the taller castor or red gram plants may be collected by hand picking or by application of suitable insecticides and thus the groundnut crop may be saved from the pest infestation to a large extent. Castor may be raised as a trap crop around chilli crop to avoid damage to the primary chilli crop from the green caterpillar *(Helicoverpa armigera)* infestation. Raising tomato as trap crop in orange orchards helps to prevent damage of orange fruits from the fruit - sucking moths *(Othreis materna* and *O. fullonica)*

c) **Mixed cropping**. This is another method by which the increase in pest population may be reduced. When some crops are sown mixed, the insects associated with one crop find it difficult to reach their natural host, since plants of other crops, which are not the hosts, are present in-between. In moving from one host to another, slow moving insects are hindered, whereas in a single crop, the pests can move easily from one host plant to another without any hindrance. Sorghum, pearl millet and red gram may be raised as mixed crop.

d) **Tillage operations**. Tillage operations such as, deep ploughing of fields, digging and turning the soil etc. help to control several insect pests, which pass one or more phases of their life cycle under the soil. In such operations the different growth stages of the pests such as, larvae and pupae are brought to the surface from under the soil, which are destroyed by sun's heat or devoured by various predators. Ploughing the soil after a few good summer showers helps to bring out the pupae of groundnut red hairy caterpillar from deep under the soil. Tilling the soil under the trees in orchards helps to bring out the pupae of fruit flies, which are found in the soil. White grubs attacking the roots of several crops such as, sugarcane, finger millet etc. can be destroyed by ploughing the soil. Ploughing and turning the soil help to bury some of the pests, which are found on the surface of the soil. Tillage operations help to destroy grasses and other weed plants in the fields, which otherwise serve as food and provide suitable environment for multiplication of many pests. Many hoppers, bugs, mealy bugs, scale insects etc. thrive on grasses and weed plants in the absence of their regular hosts. Trimming and pasting of field bunds especially in rice fields help to destroy the eggs of grasshoppers, and many bug pests.

**e) Other cultural methods**

1. Using a slightly higher seed rate and at the time of thinning and weeding, removal of infested seedlings, at the same time maintaining the normal plant population is a practice followed for the control of sorghum shoot fly and stem borer.
2. Pruning of dead branches and severely infested branches may help to control many pests. This method is followed to control stem borer and scale insects in citrus plants. Pruning of trees to avoid intensive branching and dense foliage may help to prevent severe occurrence of pests such as, mango hopper.
3. Scraping of unhealthy stems, patching up wounds with Bordeaux paste or tar and removal of loose barks from stems help to control some pests, which live or hibernate in such places in one or more stages of their life cycle. Such methods are followed to control bark borers, stem boring beetles and weevils, hoppers such as, mango hopper, which hibernate under the barks.
4. Clipping the tips of rice seedlings prior to planting helps to prevent stem borer attack to some extent, as eggs are laid mostly on the leaf tips.
5. Trash mulching on the third day after planting and earthing up on the thirtieth day after planting may prevent sugarcane early shoot borer attack, as egg laying on the young leaves is restricted.
6. Impounding water for a few days in rice nursery helps in the control of rice army worm *(Spodoptera litura)*
7. Application of balanced dose of fertilizers and proper irrigation stimulate healthy and vigorous growth of plants. Such healthy plants develop some amount of resistance and are less liable to attack by pests.

**ii) Physical methods**

By use of physical action or physical efforts some pests can be controlled to a large extent. Some insects lay eggs in masses, which are quite prominent and can easily be detected on the plant parts by the naked eye. These egg masses may be collected and destroyed. The eggs of groundnut red hairy caterpillar, rice stem borer, rice earhead bug etc. can thus be destroyed.

Newly hatched red hairy caterpillars, *Prodenia* caterpillars, semiloopers etc. are found aggregated on the under surface of leaves. These can easily be spotted and collected and destroyed. Similarly in many cases, the grown up caterpillars or adult moths can be hand picked and destroyed. Red hairy caterpillars, *Prodenia* caterpillars and semiloopers attacking groundnut, castor, bitter gourd etc., *Epilachna* beetles infesting egg plant, cucurbits etc. can be hand picked and destroyed by putting them in water covered with a film of kerosene oil or crude oil.

Some caterpillars such as, drum stick hairy caterpillars aggregate in hollowed out tree trunks and cavities in stems of trees during the day time. They can be destroyed buy using burning torches.

Coconut rhinoceros beetles can be killed by poking a long pointed and hooked needle made of steel through the bore holes. Similarly using a long and pointed needle can kill sugarcane early shoot borer.

Dragging a long rope along the top of the crop canopy from two opposite field bunds and disturbing the crop makes the rice case worms to dislodge and fall into the stagnant water, which can be collected by draining the water and then killed.

During the off seasons, burrows in the field bunds and fields can be dug and the field rats caught and killed.

Fitting toothed tin discs at right angles to the stem or tying dry thorny bushes around the stems of coconut trees at a height of 8 - 10 feet above ground level prevents rats from climbing up the trees and damage tender coconuts.

To prevent migration of insects such as, groundnut red hairy caterpillars, rice army worms, nymphs of grasshoppers etc. from infested fields to adjacent fields, trenches can be dug all round the infested fields and the insects falling into the trenches can be collected and destroyed.

Fruits such as, pomegranate can be covered with paper or polythene bags to prevent infestation by fruit borers.

Thin wire nettings may be provided to ward off mosquitoes, house flies etc. into human dwellings.

Ornamental or very rare plants may be grown in insect - proof screen houses to prevent infestation by pests.

**iii) Mechanical methods**

Several insect and non-insect pests may be caught and destroyed by the use of simple mechanical devices or gadgets. One such important method, which can be employed effectively is netting or bagging. Insects may be collected with the help of insect nets or large cloth bags by sweeping over the crop canopy against wind current and subsequently killed. This is useful to control pests such as, rice earhead bugs, sorghum earhead bugs, red pumpkin beetles, lady bird beetles etc.

Sticky traps may be used to destroy insects such as, hoppers, white flies, thrips, aphids etc.

Light traps may be set up in the fields to attract insects of several species. Moths of rice stem borer, rice green leaf hoppers, groundnut red hairy caterpillar moths, tobacco caterpillar moths, rice earhead bugs, groundnut earwigs and several other positive phototrophic insects can be destroyed by using light traps.

Different types of poison bait traps are used to attract and kill specific insect pests. Sorghum shoot flies, fruit flies, fruit - sucking moths, coconut rhinoceros beetle, red palm weevils, rats, snails and slugs etc. can be destroyed by this method.

Pheromone traps may be used to attract and kill some insect pests. Specific pheromones have been introduced to attract either the male or female insects of a species. However, most of the pheromones commonly used attract only the male insects. *Prodenia* moths, cotton pink boll worm moths, fruit flies etc. can be attracted and killed by this method.

High temperature is sometimes used for destroying insect infestation in grains such as, cereals, pulses, beans, peas etc. and other food products. No insect can survive when exposed to temperatures of 60° - 65°C. Majority of insects, including pests of stored grains and stored products are killed by continuous exposure for 3 hours at 51° - 54°C. Insects, mites and nematodes found inside bulbs, rhizomes etc. can be killed by subjecting them to heat treatment at 45°C. for 5 - 15 minutes without affecting germination.

Migrating locusts can be driven away by raising smoke screens.

Inaudible ultrasonic sound waves are used to destroy some insect pests infesting stored grains and stored products.

Ultraviolet light is used to attract and destroy many harmful insects.

Exposure of insects to radiation such as, Gamma radiation and by the use of certain chemical substances, sterility may be induced in the males of some insect species.

**iv) Use of natural agencies**

The sun is the most important natural agency, which is utilized to destroy many insect pests. Drying the seeds and grains in direct sun helps in destroying the pests present inside the seeds and by reducing the moisture content in the seeds or grains to prevent further infestation by the pests during storage.

Activated clay is another natural product, which is used to control storage pests. Activated clay is obtained by a process of treating kaolinic clay with concentrated sulphuric acid and subjecting it to very high temperature to remove all organic matter present in the clay. When seeds mixed with activated clay is stored under air-tight conditions, the clay particles absorb water from the surface of insect's body leading to dehydration and ultimate death of the insects. The treatment has to be done under air-tight conditions, otherwise activated clay will absorb moisture from the atmosphere and becomes inactive and ineffective.

Silica gel available under the trade name **'Drie-die'** is mixed with the stored grains to control the pests infesting them. The sharp edges of the silica particles cause aberrations in the epidermal layer of the body of the insects and the body fluids come out, as a result the insects are killed of dehydration. The product can also kill grasshoppers, mites, thrips, wasps etc.

**v) Legal control**

In the earlier days there were no restrictions for the movement of seeds, seed materials, plants, birds, animals etc. from one country to another. This practice has resulted in the dissemination of insects, pests and disease causing organisms to different countries leading to severe outbreaks of various pests and diseases. However, it was only since middle of the last century that serious attempts at legislation to prevent the spread of pests and diseases were enacted by different countries. The main aim of such legislations are to prevent the introduction of foreign pests and diseases from one country to another, to prevent or restrict spread of serious crop pests and diseases from one country to another, to prevent or restrict spread of serious crop pests and diseases within a country and to ensure the quality of plant protection chemicals. Cottony cushion scale, wooly aphid, San Jose scale, potato golden nematode and giant African snail are a few pests, which had been introduced to this country from foreign countries. Some of the very serious pests such as, Mediterranean fruit fly, Mexican cotton boll weevil and codling moth, which are found in other countries, have not found their way to our country so far.Prohibitive laws to prevent introduction or spread of any dangerous pest or disease are called **'quarantine laws'**. In India **'The destructive insects and pests Act of 1914'** was enacted by the

Government in 1914 to prevent introduction of destructive pests and diseases from foreign countries to India.

The quarantine laws are implemented in 5 different ways

a) **Laws to prevent entry of pests, diseases, seeds etc. from foreign countries.** 'The destructive insects and pests Act of 1914' was enacted in 1914 in India to achieve this objective. According to this Act, seeds, seed materials, plants etc. brought to India should be accompanied by a **'phytosanitary certificate'** from the authorized official of the exporting company to the effect that the materials are free from pests or any disease causing organisms. On entry, these materials are thoroughly examined for the presence of insects, diseases, weed seeds etc. and only when such materials are found to be free from pests etc., they are given proper treatment and then allowed into the country, otherwise they are destroyed. The quarantine centres, which are located at every point of entry viz', Airports of Amritsar, Mumbai, Calcutta, Chennai and Delhi and harbours at Mumbai, Calcutta, Chennai, Cochin, Tuticorin, Rameswaram, Bhavanagar and Vizagapattinam are responsible for testing the materials received and permit entry only when they are free from pests etc. Import of any seeds through post is totally prohibited. Only some scientists are permitted to import small quantities of seeds for research purposes.

   As per the Act, import of potato is prohibited from countries where **'wart disease'** and **'golden nematodes'** are prevalent. Further, import of rubber seeds, sugarcane setts, coffee seeds, cotton seeds etc. is prohibited.

   Similarly pepper, tamarind, cardamom etc. exported to foreign countries from India should also be accompanied by phytosanitary certificates from concerned officials.

   The quarantine centres established in the various airports and harbours function under the **'Central Directorate of Plant Protection and Quarantine'**, which was established in 1946. From 1949 the Directorate has established quarantine stations in a number of sea and airports and land frontiers to monitor and carry out these activities.

b) **To prevent spread of already existing pests, diseases etc.** According to the 'Destructive insects and pests Act of 1914', every state in the country is empowered to enact their own Acts to establish domestic quarantines to prevent interstate and intrastate spread of pests and diseases. To achieve this objective, the Tamil Nadu Government had enacted **'The Madras Agricultural Pests and Diseases Act of**

**1919'**. In the year 1919, the Act was passed with a view to prevent introduction, spread or reappearance of pests and diseases. According to this Act, the Government can declare a pest as epidemic and invoke the Act and direct the Agricultural Department to take appropriate control measures such as, **'Mass ground spraying'** or **'Aerial spraying'**. When an epidemic of 'cottony cushion scale' appeared in the Nilgiris and Kodaikanal regions, this Act was invoked in 1943. Transportation of the host, as well as the alternate hosts of this pest from these regions to other parts of the country was prohibited. Quarantine stations were opened in 1943 at Mettupalayam, Gudalur, Shenbaganur and Top Station for necessary enforcement of this Act.

c) **Act to make obligatory control of epidemic pests**. To prevent further spread of established pests or diseases and to avoid further damage when there is an epidemic outbreak, **'The pest and disease Act'** is invoked by the Government as and when necessary to make it obligatory on the part of the farmer to adopt control measures recommended by the Department of Agriculture as a joint venture.

For the control of cotton boll worms and stem weevil, **'The cotton pest Act'** was enforced in 1918 in some parts of Tamil Nadu. For the control of black-headed caterpillar of coconut, the Act was introduced in 1927 and 1934 in some parts of Tamil Nadu. In 1929, the Act was introduced in the Nilgiris to prevent the spread of cottony cushion scale *(Icerya purchasi)* to other parts of the hilly region. Import of coffee seeds from other countries was banned from 1927 to prevent introduction of coffee berry borer . Similarly for the control of groundnut red hairy caterpillar and the coffee stem borer, **'The Madras Agricultural pests and disease Act'** was invoked in 1930 and 1946 respectively in some parts of Tamil Nadu. In 1958, the Act was enforced in some pars of Tamil Nadu sugarcane belts to prevent the spread of sugarcane top borer by which all top borer attacked shoots of sugarcane had to be removed and destroyed. To prevent the spread of the cattle fly *(Stomoxys calcitrans)*, the Act was enforced in 1943 in parts of Tamil Nadu by which application of groundnut cake as manure to crops was prohibited, as the fly was found to breed and multiply in groundnut cake applied as manure to crops. In Uttar Pradesh, **'The locust destruction Act, 1951'** was promulgated in 1951 and in 1954 for the control of migratory pests.

d) **Quality control of pesticides**. To prevent adulteration of plant protection chemicals, formulation of sub-standard chemicals and to prevent furnishing of false trade informations about such chemicals, an Act called **'The insecticide Act of 1968'** was promulgated in 1968. According to this Act, the companies manufacturing or formulating plant protection chemicals should register their company and brand name with the Government Agency. They should furnish the technical name , trade name and percentage of active ingredient, net content in the package, method of application and dosage, antidotes and all such informations on the labels of the container. Pesticides, which are highly toxic to man, animals and poultry are required to carry on the label skull and cross bones symbol indicating a warning. Further, the label should have **'red'**, **'yellow'**, **'blue'** and **'green'** markings indicating nature of the poison viz., 'very highly toxic', 'highly toxic', 'moderately toxic' and 'less toxic' respectively. Quality control tests of pesticides have also been made mandatory.

**'The Insecticides Act - 1968'** also envisages restrictions on the import and formulation of plant protection chemicals, transporting of such chemicals within the country and marketing, so as to prevent health hazards to man and other livestock due to pesticides. **'The Central Insecticide Board'** constituted by the Government of India in 1971 monitors the implementation and functioning of this Act and advices the Central and State Governments on technical matters.

e) **Act to stress precautionary measures to be adopted while handling pesticides**. Most of the plant protection chemicals are highly toxic to human beings and other livestock. The toxic substances may enter into the body through the mouth, nose or skin and cause dangerous consequences. Hence, great care should be taken while applying such chemicals either by way of spraying, dusting, fumigating, soil injecting or by any other methods. To enforce adoption of certain protective measures while handling pesticides, some rules and regulations have been formulated.

### vi) Biological control

The introduction, encouragement and artificial increase of predacious and parasitic insects (Parasitoids), animals, birds and disease causing organisms for the control of injurious insect pests is called **'Biological control'**. Though it is related to normal control to some extent, man plays an important role in this method of control. Several parasitic and predacious insects, predatory vertebrates, parasitic fungi, bacteria, viruses, nematodes, protozoa etc. serve as agents of biological control.

Frequent and indiscriminate use of pesticides has resulted not only in the development of resistance in insects and mites, but also in the destruction of their natural enemies leading to several pest outbreaks. Besides this, these synthetic chemicals are highly toxic and are hazardous to human and animal lives. They also pollute the air, soil and water and cause dangerous after effects. All these factors have been responsible for inciting considerable interest in the adoption of biological control in the recent years.

### a) Parasitoids

**'Parasitoids'** are entomophagous insects, which live on or in the body of another insect known as **'the host'** from which they get protection and nourishment at least during one growth stage of their life cycle. In a typical case, the parasitoid lays eggs on or in the body of the host. The larvae of the parasitoid emerging from the eggs feed on the body contents of the host and emerge as adults. The hosts are not killed immediately but continue to live for a rather a long period, at least till the larvae become full - grown.

Parasites that live on the body of the host are called **'ectoparasites'**, while those that enter and live inside the body of the hosts are called **'endoparasites'**. If a parasitoid is capable of attacking a single species of insect , it is called **'monophagous'**. If it is capable of attacking a few closely related insect species, it is called **'oligophagous'** and if it can attack a number of widely different species of insects, it is known as **'polyphagous'**. In most cases only one parasitoid emerges from the host after completing its life cycle and such parasitoid is called a **'solitary parasitoid'**, while in a few other cases more than one parasitoid emerge from the host and these are called **'gregarious parasitoids'**

Among the parasitoids up to about 80 per cent belong to the Order-Hymenoptera and Diptera and are generally known as **'wasp parasitoids'** and **fly parasitoids'** respectively. Parasitoids are mostly very small in size and some of them can be seen only through a microscope. They have four different growth stages viz., egg, larva, pupa and adult and in the larval stage alone they live as parasites on the host. In general, the parasitoids attack any one particular growth stage of the host only.

Parasitoids by their sense of smell, touch, feel of the ovipositor, as well as the odor emanated from the saliva and excretory matter of the host, find and reach the hosts and lay eggs on the body or inside the body of the hosts

A parasitoid to be successful, should have the following qualities: It should be adaptable to the environmental conditions of the new locality and should be able to survive under the new conditions, which is occupied by the host pest; It should be specific to a particular species of the host or at least should have a narrow and limited host range; It should be able to multiply faster than the host, have short life cycle and high female-male ratio; It should have sense of finding the host and should be amenable for easy mass breeding under controlled conditions; It should be able to disperse easily and quickly in the new surroundings; It should be free from attack by hyper parasitoids and predators; Its rate of multiplication and development should synchronize with the vulnerable growth stage of the host.

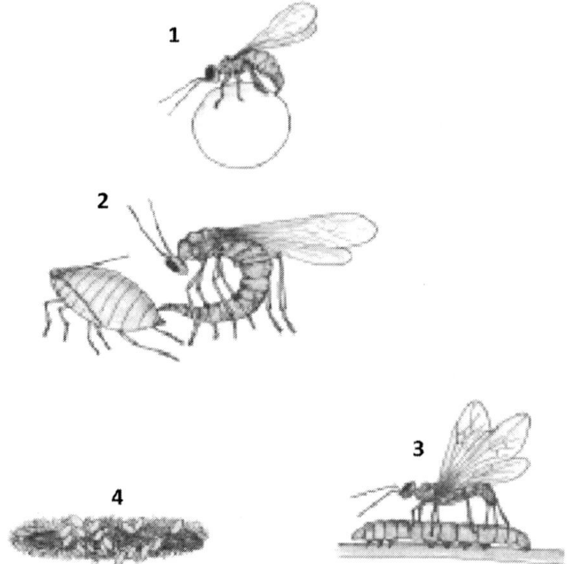


**Fig. 1.** Parasitoids attacking different growth stages of pests

1. Trichogramma egg parasitoid attacking an egg of its host 2. Eulophid attacking an aphid 3. Braconid attacking a larva of its host 4. More than one pupae of the parasitoid on the host (Gregarious parasitoid)

1. **Egg parasitoids**. Among the parasitoids, species of *Trichogramma, Telenomus* and *Tetrastichus* are most important. The genera *Trichogramma*, which belongs to the Family - Trichogrammatidae under the Order - Hymenoptera consists of very small, wasp-like

insects, about 0.5 mm. in size, brownish in color with small wings. Their life cycle is completed in about 7 days inside the host eggs and hence several generations may appear within a short period of time. Species of *Trichogramma* parasitize and destroy eggs of sugarcane early shoot borer, sugarcane internode borer, cotton boll worms, stem borers of rice and sorghum, shoot borers of egg plant, tomato etc., To control sugarcane internode borer, *Trichogramma* egg parasitoids have to be released six times at 15 days intervals from the fourth month after planting at 16,000 - 18,000 parasitoids per acre per release.

Egg parasitoids of the genus *Telenomus* and *Tetrastichus* parasitize the eggs of stem borers of sugarcane, rice, pulses, cotton etc. These parasitoids are often found in the fields naturally.

Some of the important egg parasitoids and their respective hosts are presented in **Appendix - 3**

2. **Egg - larval parasitoids**. This type of parasitoids initially attack the eggs of the hosts and subsequently, when the eggs of the host hatch, the parasitoids parasitize the larvae and ultimately kill them. *Chelonus blackburni* (Order - Hymenoptera, Family - Braconidae) parasitize the eggs and then the larvae of cotton spotted boll worms, cotton pink boll worms, potato tuber borers etc. These parasitoids are slightly bigger in size and black in color with yellow bands on the abdomen. They have to be released at the rate of 16,000 adults per acre at 15 days intervals after the pest occurrence is noticed,

3. **Larval parasitoids**. The larval parasitoids parasitize only the larval stage of the host,. These parasitoids commonly called as Braconids, Bethylids and Ichneumonids belong to the Order - Hymenoptera and are wasp type of parasitoids, while the ones commonly called as Tachinids belong to the Order - Diptera and are fly type parasitoids.

   **Braconids**. These parasitoids are small, delicate insects, brown in color and appear as red ants. Their life cycle is completed in 10 - 14 days. They are released at the rate of 1,600 - 2,000 insects per acre at 15 days intervals after the pest infestation is noticed. They parasitize cotton spotted boll worms, cotton pink boll worms, green caterpillars, *Prodenia* caterpillars, groundnut leaf miner, coconut black - headed caterpillars, fruit borers of vegetables, stem borer of rice, sugarcane, sorghum etc.

   **Bethylids**. These parasitoids are small, ant - like, shining, black - colored insects. They parasitize the larval stage of groundnut leaf miner, semiloopers, cotton green caterpillars, *Prodenia* caterpillars, coconut black - headed caterpillars etc. To control coconut black - headed caterpillar, the parasitoids are released at the rate of 10 insects per tree along with their pupal parasitoids.

**Chalcids**. These are very small insects. The abdomen is raised like a hump. The femora of the hind legs are very stout. The ovipositor is straight and long. They parasitize the larvae of many insects belonging to Lepidoptera, Diptera and Coleoptera.

**Ichneumonids**. These insects are slightly bigger in size. The antennae and legs are well developed. With their well developed, strong and long ovipositor, they penetrate the stem of plants and lay eggs on the larvae hiding inside the stem. Internal feeders such as, stem borers, leaf miners, leaf folders etc. are attacked. Their life cycle is completed in 18 - 23 days. Coconut black - headed caterpillars, Sugarcane top borer etc. are controlled by these parasitoids.

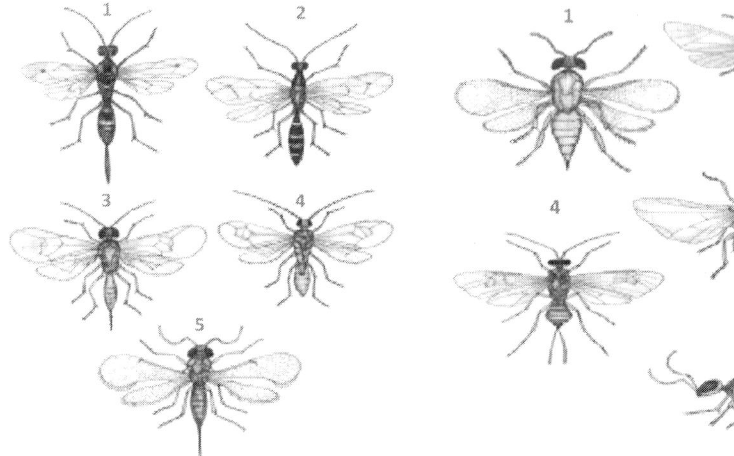

**Fig. 2.** Types of parasitoids 1. Female Ichneumonid sp. 2. Male Ichneumonid sp. 3. Female Braconid sp. 4. Male Braconid sp. 5. Female Trichogramma sp.

**Fig. 3.** Types of parasitoids 1. Female Eulophid sp. 2. Chalcid sp. 3. Tachinid sp 4. Microbracon sp. 5. Bethylid sp.

**Elasmids.** They are very small, wasp - like ectoparasites. Their life cycle is completed in 10 - 16 days. They parasitize coconut black - headed caterpillars etc.

**Tachinids**. Tachinids are solitary insects. Only one parasitoid emerges from each host. They are slightly bigger in size and are fast fliers. Their life cycle is completed in 18 - 24 days. The eggs of the parasitoid hatch within the genital chamber of the mother and when the young larvae are able to crawl about a little, the mother deposits them near the locations where the host larvae are present. The parasitoid larva crawls slowly, reaches the host larva and penetrates and enters. inside. The full-grown parasitoid larva comes out of the host and pupates on the surface of the host. The pupa is seed \- like and red in color. Species of *Sturmiopsis* are the most important ones. The parasitoids are released at the rate of 5 female insects per acre at 15 days intervals.

Some of the important larval parasitoids and their respective hosts are furnished in **Appendix - 4**

**Pupal parasitoids**. Parasitoids belonging to Eulophids and Ichneumonids are the most important pupal parasitoids. They attack the pupae of *Prodenia* caterpillar, American cotton boll worm, castor capsule borer, coconut black - headedcaterpillar, leaf folders of cotton, rice, snake gourd etc. The young pupal stage is more vulnerable to attack by the parasitoids. From each pupa of the host 30 - 50 parasitoids may emerge.

Some of the most important pupal parasitoids are listed in **Appendix - 5**

**b) Disease causing microorganisms** or **Parasites**

As in the case of parasitoids, which cause slow death of their hosts, several parasitic microorganisms such as, fungi, bacteria, viruses, protozoa and nematodes also infest insects, live on the body contents of the hosts, multiply enormously on or in the body of the insects, cause diseases and ultimately destroy them.

Among the fungal parasites, *Metarrrhizium anisopliae,* commonly known as the **'green fungus'** infests the larvae, as well as the adults of coconut rhinoceros beetle and destroy them. This fungus can be multiplied in substrates such as, carrot, tapioca etc. and incorporated in the manure pits to destroy the grubs. The fungus can also be multiplied in the grubs of rhinoceros beetle.

Another fungus, *Beauveria bassiana,* commonly known as the **'white fungus'** infests and kills insects such as, rice brown plant hopper, rice green leaf hopper, sugar beet borer etc.

The fungal parasites function effectively only under conditions of high humidity and moderate temperature. So, during the rainy seasons they can be used effectively.

Some of the important fungal parasites and their specific hosts are furnished in **Appendix - 6**

Some bacterial species are also parasitic on some particular insects. Among them the most important one is *Bacillus thuringiencis*, which is capable of killing the larvae of many crop pests. The spores of this bacterium in suitable media are available under the trade name **'Thuricide'**, **'Delfin'**, **'Difel' etc**. The bacteria are capable of producing a toxic protein metabolite in the intestine of the hosts and thereby destroy them. There are a few other bacterial species, which also can destroy particular insects.

Some of the important bacterial parasites are furnished in **Appendix - 6**

A few viruses also play an important role as parasites of crop pests. *Nuclear polyhedrosis viruses* and *Granulosis viruses* help to control some of the very important crop pests effectively. *Nuclear polyhedrosis virus* to control the larvae of pests viz., groundnut red hairy caterpillar, *Prodenia, Heliothis, Agrotis* etc. are being employed on field scales. Specific viruses have to be used to control particular species of insects. Commonly called as **'NPV'**, these viruses can be multiplied only in the hosts, which they parasitize and are more effective in destroying the early instar larvae, which are active feeders.

The *Granulosis viruses* can control sugarcane early shoot borer, rice leaf folder, cabbage semilooper, apple fruit borer etc. *Baculo virus,* another virus can attack and destroy the grubs, as well as the adult coconut rhinoceros beetles.

A few nematode parasites have also been found to parasitize and destroy some specific insects. Some species of *Rhabditis* can destroy larvae of sorghum stem borer. Spraying of a culture of the nematode *Neoplectana carpocapse* is being followed against various insect pests of economic importance such as, rice and sugarcane borers and codling moth.

Similarly some protozoa are also parasitic on some particular insects. Some species of *Tetrahymena* can destroy the larvae of sorghum stem borer.

**c) Predators**

Predators include insectivorous insects, non - insects such as, spiders, vertebrates such as, birds, reptiles, frogs and toads etc., which catch and feed on insects by devouring them immediately.

A total of about 220 species of insects, which are predators are found throughout the world. Among them, some beetles, bugs, wasps, lace wing bugs, dragon flies, damsel flies, praying mantis, flies etc. are most important. Predatory insects are generally larger in size than their preys and are very active. Their body organs are suitably modified for catching the prey and feeding on them.

## Predatory insects beneficial in crop protection

**Lady bird beetles**. The grubs and adult beetles of several lady bird beetles belonging to the Order - Coleoptera are predatory on aphids, leaf thrips, mealy bugs, scale insects, hoppers etc. Species of *Coccinella* and *Brumus* are predatory on many small insects. Grubs and beetles of *Coccinella septum punctata* prey on sorghum stem borer, sugarcane leaf hoppers etc. *Coccinella undicempunctata* is also predacious on sugarcane leaf hoppers. *Coccinella sexmaculata* feeds on coffee scale insects. Several species of *Brumus* destroy cotton white flies. *Brumus suturalis* destroys sugarcane leaf hoppers. *Brumoides suturalis* is predacious on sorghum stem borer. *Cryptolaemus montrouzieri* feeds on scale insects and mealy bugs. *Chilocorus nigritus* preys on coconut scale insects and coffee scale insects. *Rodolia cardinalis* is predacious on citrus scale insects **(Fig. 4)**

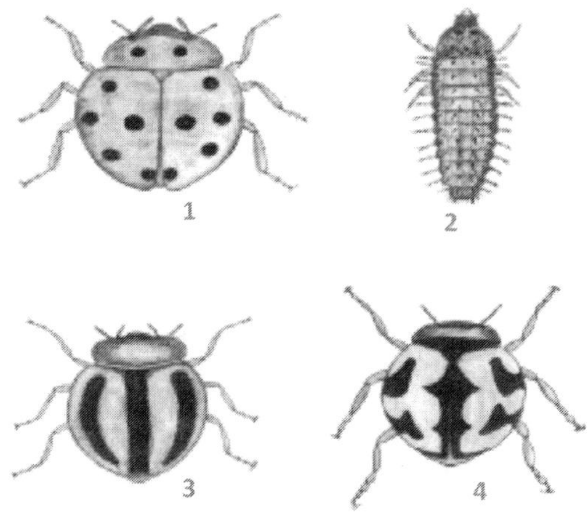


**Fig. 4.** Types of predatory insects - Lady bird beetles

1. Adult beetle - *Coccinella* 2. Larva 3. *Brumus* species 4. *Rodolia* species

**Dragonflies and Damsel flies**. These predators, which belong to the Order - Odonata feed on mosquitoes, flies, moths, winged termites and many other insects while in flight. They are fast fliers, with incurved, basket - like thorned legs and large labrum for catching insects in flight. The naiads, which live in water feed on insects living in water.

**Praying mantis**. These insects belonging to Order - Dictyoptera are predatory on many insects. Both the nymphs and adults prey on flies, grasshoppers, larvae etc. Their fore legs are adapted to catch and kill insects.

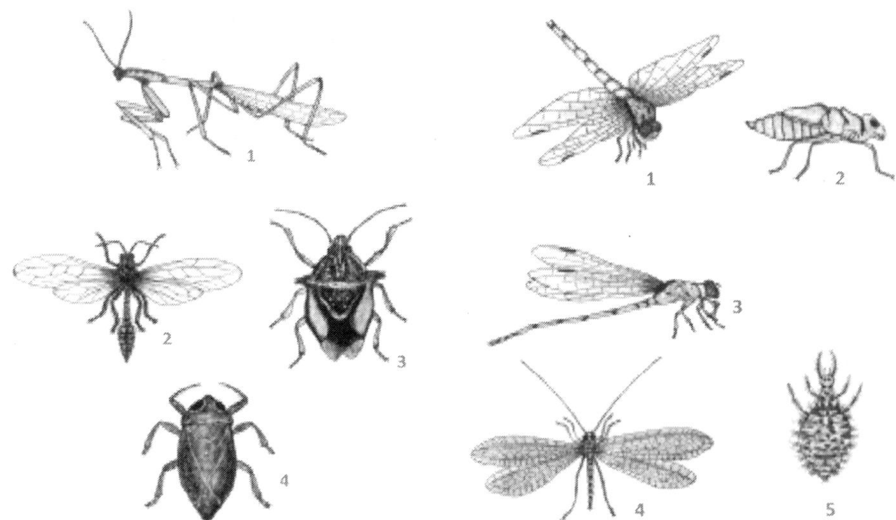


**Fig. 5.** Types of predatory insects
1. Praying mantis *(Mantis religiosa)*
2. Wasp *(Eumanes petiolata)*
3. Shield bug *Canthecona furcellata,*
4. Giant water bug *(Belostoma indicum)*

**Fig. 6.** Types of predatory insects
1. Dragon fly (Adult) 2. Dragon fly naiad
3. Damsel fly 4. Green lace wing bug
5. Ant lion larva

**Green lace wing bugs.** The larvae of these insects belonging to Order - Neuroptera feed on aphids, scale insects, mealy bugs, hoppers, leaf thrips, mites, white flies etc. The mouthparts of the larvae, which are long and needle - like, pierce the body of their prey, suck and feed on the inner contents and destroy them. One such predatory larva can destroy 400-500 insects. *Chrysoperla cornea,* the green lace wing bug is predatory on a number of insect pests infesting cotton, egg plant, groundnut, fruit trees etc.

**Wasps**. Some wasps belonging to the Order - Hymenoptera catch larvae of many insect pests and use them as food for their own grubs. Wasps such as, the solitary mud wasps and digger wasps construct nests with many chambers and place one larva, which has been made unconscious by its sting in each of the chamber, lay an egg on each larva and then close the chambers. The grubs, which hatch out of the eggs feed on the host larvae and destroy them. In each nest there may be 10 - 12 host larvae.

The boll and pod borer pests of cotton, red gram, beans etc. are thus destroyed by wasps. The wasp - *Eumanus petiolata* destroys the larvae of cotton boll worm pest in this way.

**Ants**. Some species of ants belonging to the Order - Hymenoptera are predacious on a number of insects. The tree ants - *Oecophylla smaragdina* web several leaves together and build box - like nests and live inside the nests in trees such as, mango, cashew, guava etc. These ants feed on several insects, as well as eggs, larvae and pupae of other insects.

**Ant lions**. The larvae of this insects belonging to Order - Neuroptera make sloping pits in loose soils and remain hidden inside the bottom of the pits. Insects, which move on the soil surface skid and fall into the pits and are caught and eaten by the larvae. The larvae of *Chrysopa cymbela*, also of the Order - Neuroptera feed on cotton white flies.

**Syrphids (Hover flies)**. The larvae of several Syrphids belonging to the Order - Diptera feed exclusively on aphids, coccids and other small Homopterans. The larvae with their pointed mouthparts pierce, suck and feed on the inner contents of aphids and other hosts. A single larva may destroy more than 400 aphids. The species belonging to *Syrphus* and *Parachus* are important predators.

**Ground beetles**. These beetles belonging to the Order - Coleoptera have long legs and are fast runners. The beetles, as well as the grubs are predacious. The beetle - *Callida splendidula* and its larvae feed on coconut black - headed caterpillars. One insect may destroy up to 8 black - headed caterpillars in one day. The beetles - *Ophiomyia* sp. are bright red in color, with black bands across. They are mostly found in rice fields and feed on brown plant hopper, leaf hoppers, leaf folders etc.

**Pentatomid bugs** or **Shield bugs**. These bugs belonging to the Order - Hemiptera feed on many insects. They have long, needle - like mouthparts with which they pierce the body of their prey, feed on the inner content and destroy them. The green mirid shield bug - *Cyrtoragins lividipennis* feed on rice brown plant hopper. The assassin bug - *Platymeris* sp. attacks and kills even coconut rhinoceros beetle. *Canthecona furcellata* feed on groundnut red hairy caterpillars. *Andrallus spinidens* feed on *Prodenia* caterpillars.

Some of the bugs belonging to Hemiptera, which live in water such as, the water straddlers, water skaters, giant water bugs etc. are also predatory in nature. The fore legs of water straddlers are short and spaced much in front of the hind legs. The body and the legs are very slender like sticks. They have piercing and sucking type of mouthparts. They feed on several rice pests such as, brown plant hoppers, green leaf hoppers, leaf folders, case worms, stem borers, army worms etc. They are capable of running very fast on the surface of water, catch and feed on these pests. One insect may destroy 2 - 4 insects in one day. The water skaters are small insects, 3 - 4 mm. in length. They are found on the surface of water and under the plant canopy. Both winged and apterous forms may be present. They suck and feed on the inner contents of brown plant hoppers and leaf hoppers. One insect may feed on 3 - 6 pests in one day

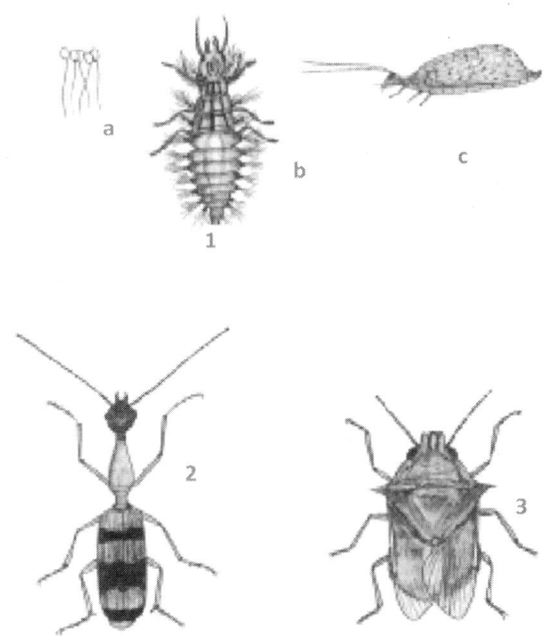


**Fig. 7.** Types of predatory insects
1. Green lace wing bug (a) Eggs (b)Larva (c) Adult bug
2. Ground beetle *(Ophiomyia sp.)* 3. Predatory stink bug *(Andrallus sp.)*

**Non - insect predators**. A few species of spiders, which are usually found in rice fields and other crop fields, as well as frogs and toads, fishes, ducks, crows, king crows, mynahs, storks, cranes, owls, wood peckers etc. also feed on insects.

Among the predatory spiders there are two types viz., the web spinning type and the non - web spinning type. The web - spinning spiders viz., the 'Long - jawed spiders', 'Dwarf spiders' and 'Round spiders' are the important ones. They spin webs in the foliage of crops and remain in the centre of the webs. Flying insects, which get trapped in the web are caught by the spider and made unconscious by biting. Then they suck and feed on the body contents leisurely. Moths, leaf hoppers, brown plant hoppers etc. are destroyed by them.

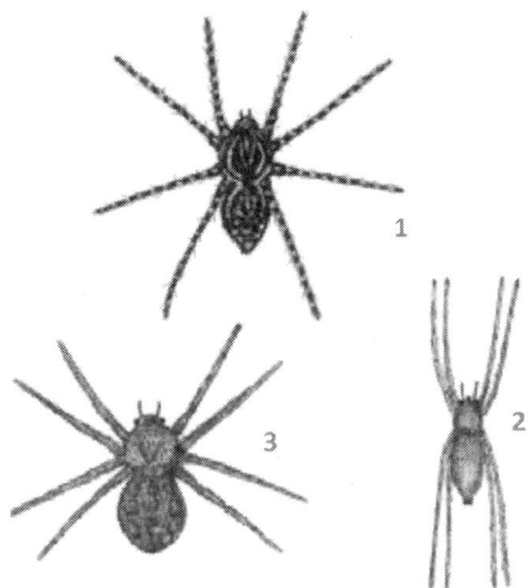


**Fig. 8.** Types of predatory spiders
1. Wolf spider 2. Links spider 3. Short spider

Wolf spiders, Leaping spiders and Links spiders are important among the non - web spinning type. These spiders usually remain on the basal portions of plants. The wolf spider - *Lycosa pseudoannulata* is commonly found in rice fields. They are fast runners and they attack, catch and prey on several insect pests. One spider may destroy 10 - 15 insects in one day. They are also capable of swimming in water and hence can catch and feed on insects living in water too. They destroy insects such as, stem borers, leaf folders, brown plant hoppers, leaf hoppers etc.

The links spider - *Aclyopus javanus* is also found in rice fields. They usually hide behind plant parts and suddenly jump on the prey, catch and feed on them. Moths, leaf hoppers, brown plant hopper, green leaf hopper and other insects are their common prey. One spider may destroy 2 - 3 moths in one day.

## Methods of implementing biological control

Biological control is implemented in 4 ways :

1. Parasitoids and predators may be collected from places where they are found in large numbers under natural conditions and introduced afresh in places, where they are not found naturally even when there is occurrence of their hosts.
2. The host pests parasitized by the parasitoids under natural conditions are collected and kept safely and when the adult parasitoids start emerging, they may be released in new localities where the pests are found.

3. The parasitoids and predators may be multiplied in large scale under artificial conditions and released when there is occurrence of their hosts.
4. Exotic parasitoids, predators and disease causing fungal, bacterial, viral and other parasites may be imported from foreign countries and introduced in localities where the host pests are prevalent.

Though parasitoids and predatory insects are entomophagous insects, there are characteristic differences between the two in their morphology, habit, nature of attack etc., which are furnished in **Appendix - 7.**

## Advantages of biological control

1. The biological control agents can operate on large areas and parasitize their hosts
2. Though introduction of biological control agents is very difficult and may incur lot of expenditure, in due course, when reintroduction becomes unnecessary, a lot of labour and expenditure can be saved.
3. When it is not possible to control certain pests by other methods of control, the parasitoids are capable of reaching them in their protected hiding places, parasitize and destroy them.
4. As long as the hosts are present, the biological control agents also multiply naturally and destroy their hosts.
5. Biological control has its permanent effect. Once the biological control agents become well established in a place, they may persist in that place permanently and further release of these agents may not be necessary.
6. Biological control agents are non - poisonous and hence they do not have any toxic residual effect on the plants.
7. Beneficial insects do not develop any adverse effect either on the crops or on the environment as in the case of plant protection chemicals.
8. Some of the biological control agents can be multiplied on a large scale under artificial conditions cheaply and without much difficulty.
9. Insects never become resistant against biological control agents
10. A few biological control agents can be deployed along with insecticides, so that the pests can be destroyed by both insecticides and the biological control agents.

## Disadvantages of biological control

1. Biological control is a slow progressing method and may take a considerably long time, sometimes many years to make a biological agent to establish in a particular place permanently.

2. Biological control agents are slow in action and hence it requires sufficient time for the destruction of the target hosts.
3. Biological control agents have narrow spectrum, as they kill insects of a particular species only. Hence these agents cannot be deployed exclusively to destroy all species of pests of a particular crop or pests in a particular place.
4. When a biological agent has several alternate hosts, it is not quite useful to control a particular pest.
5. If it is not possible to establish a biological agent in a place permanently, then repeated release of the agent becomes necessary.
6. When there are hyper parasites, which may parasitize the biological control agent, then their efficacy is nullified.
7. Several insecticides are harmful to the biological control agents and may adversely affect their existence.
8. Some of these beneficial insects are effective only under particular environmental conditions.
9. Exotic biological control agents can survive and be effective only if the climatic conditions are favorable for their existence in the new location.
10. Biological control involves living organisms and hence qualified persons are required for its implementation.
11. Biological control involves lot of labour, time and expenditure.

**vii) Chemical control**

When preventive measures of control, as well as cultural, physical and mechanical methods of control are ineffective and when the pest population and the pest damage reaches the economic threshold level, it becomes necessary to adopt chemical control measures. Under such circumstances certain chemicals are applied to prevent further pest infestation or to destroy them to avoid further damage to crops. These chemicals are known as **'pesticides'**. To meet the ever increasing demands of man, to fulfill his requirements of food, clothing and shelter intensive methods of Agriculture have to be adopted to increase the yield and to achieve this, use of pesticides to control the pests becomes imperative. Further, when there is sudden, large scale occurrence of pests , use of pesticides to eliminate them quickly seems to be the only way.

Up to a few decades back, certain plant products and some inorganic chemicals alone were used as pesticides. However, after invention of

DDT in 1939, a revolution had started in the field of plant protection. Several plant protection chemicals followed introduction of DDT in succession during the coming years.

Selection of plant protection chemicals depends upon the crop on which it has to be applied, the growth stage of the crop, the insects, which can be controlled by the chemicals, the growth stage of the pest, its habitat and nature of feeding, the effect of the chemicals on beneficial insects etc.

The details about the various plant protection chemicals commonly used have been dealt separately under the head 'Pesticides' (Page 71)

**viii) Other modern methods**

Some recent and modern methods have been introduced for the control of various pests without adversely affecting the environment. Most of these methods, though not capable of attacking and killing the pests directly, are used to cause their death indirectly. These methods include use of various attractants, repellants, sterilants, hormones, radiation etc.

1. **Radiation**. When insects are subjected to certain quantity of a particular radiation, it may induce sterility in insects. The sterility may be due to inactivation of the sperms of the male insects or due to destruction of the tissues of the Genital organs of the male and female insects or the insects may be incapacitated from undergoing the act of mating, leading to failure of the insects to reproduce. The male insects, which have been made sterile by subjecting to radiation, even if they mate with females, the eggs laid by them are abnormal. Larvae or nymphs will not emerge from such eggs. Even if young ones emerge, they are quite under developed and die soon.

   Exposure of grains to gamma radiation at 10 to 20 kilorads is effective in sterilization and subsequent kill of storage pests. However, this method is not advisable for the control of pests in grains intended for seed purposes, as it may adversely affect the germination of the seeds.

   The beetle pest - *Anobium punctatum*, which damage furniture and the powder post beetle - *Lyctus brunneus*, which damages wooden structures etc. can be completely sterilized when subjected to gamma radiation at 8.0 kilorads

   This method is applied for the control of Mediterranean fruit fly - *Ceratitis capitata,* cucurbit fruit fly - *Dacus cucurbitae*, mango fruit fly - *Dacus dorsalis,* white grub attacking the roots of vegetable crops - *Melolontha vulgaris,* mosquito - *Culex pipens fatigans* etc..

2. **Chemosterilants**. Certain chemicals when sprayed on the insects or when fed to insects or when given through tropical injection induce sterility, but do not affect their lives or their ability to mate. Different chemicals are available to make either the males or the females or both the males and females sterile. Chemicals such as, Amethopterin, Methotrexate, Fluorouracil etc. can induce sterility in insects. Chemicals such as, Triphenyl tin hydroxide (Du-ter), a potential fungicide, Triphenyl tin sulphide etc. induce sterility in both males and females of some insect species. The application of such sterilants and the period required to induce complete sterility are very important. Continued exposure to sub-sterilizing dosages of chemosterilants may lead to development of resistance in insects exposed to such dosages and certain other undesirable effects.
3. **Antibiotics and sulfanilamides**. A few of these substances, which are capable of attacking and killing several disease causing microorganisms such as, bacteria and mycoplasma are capable of destroying some insects directly. These substances when fed to insects destroy the microorganisms such as, certain bacteria, yeast and protozoa, which live securely as symbionts inside the stomach and fat bodies of insects. These microorganisms, which assist in the digestion of food and in bringing about certain biochemical changes in insects, thus help them in their life activities. When these microorganisms are destroyed by the antibiotics and sulfanilamides, the growth of the insects are drastically affected and their genitals shrink, as a result they become sterile due to atrophy of ovary and damage to follicular epithelium. The antibiotic **'Anthromycin'**, when introduced into the body of the fly - *Drosophila melanogaster* induces sterility. Antibiotics such as, Cytovirin, Cycloheximide, Streptovitacin, Antimycin-A, Hygromycin-B etc. can destroy the aphids - *Myzus persicae, Aphis pomi, Macrosiphum pisi* etc. by inducing sterility. When the antibiotic **'Terramycin'** is sprayed on the nymphs of *Aphis fabae*, sterility is induced. A few antibiotics such as, Flavomycin, Terramycin, Aureomycin etc. induce sterility in the two-spotted spider mite - *Tetranychus urticae* and the European red mite - *Panonychus ulmi.* The sulfanilamide - **'Likuden'** induces sterility in the aphids - *Aphis fabae.* The Antibiotics viz., Terramycin, Aureomycin etc., as well as a few sulfanilamides adversely affect the egg laying capacity of the red flour beetle - *Tribolium castaneum* to a considerable extent.

4. **Hormones**. Some synthetic chemical substances act as hormones, juvenile hormones or moulting hormones. These chemicals, which control the growth and development of insects, also known as **'mimics'**, besides affecting the growth of insects, induces sterility or destroy them before their growth is completed. Some growth hormones, when applied to plants develop certain morphological and chemical changes and the insects feeding on such treated plants develop sterility. **'Gibberellic acid'**, which is applied to plants as a growth hormone, adversely affects the reproductive capacity of the mites - *Tetranychus telarius,* which infest bean, *Panonychus ulmi*, which infests apple, as well as the polyphagous *Prodenia* caterpillar pest. **'Cycocel'**, another growth hormone affects the growth of the aphid - *Myzus persicae*. **'Maleic hydrazide'**, another growth hormone affects the growth and induces sterility in the aphid - *Acyrthosiphon pisum.* The weedicides **'Amitrole'**, **Zytron'** etc. also induce sterility in the aphid - *Acyrthosiphon pisum* infesting peas.

   Growth inhibitors or juvenile hormones are also used for the control of some insects. These substance called **'Ecdysones'** obtained from some Arthropods and some ferns, when sprayed on the insects or fed to insects develop some changes in their growth and feeding habits and ultimately kill them. This substance when given to the red cotton bug - *Dysdercus koenigii* in small quantities inhibits the insects from feeding and when fed in large quantities induces sterility.

   Some of the juvenile hormones, when sprayed on the pupae, the adults that emerge from the pupae are under-developed and die soon.

   **'Juvabione'**, **'Dehydro juvabione'**, **'JH-1'**, **'JH-2'** etc. possess such qualities. The juvenile hormone **'Altocid SR-10'**, when mixed with water and sprayed controls mosquitoes. **'Altozar'** and **'Altocid'**, when mixed with food in small quantities and fed to storage pests control them effectively. A few chemical compounds belonging to the **'Peptides'** and **Terpenoids'** possess high degree of JH activity. Use of JH analogues in pest control is receiving considerable interest in recent years.

5. **Genetic manipulation**. Manipulation of genes also develops some changes in the body of the insects and induces sterility. Some of the deleterious genes present in insects can be introduced into the existing pest population, either to induce sterility or alter the sex ratio or to produce other harmful effects and thus the pest can ultimately be destroyed. This method is being used to destroy mosquitoes.

6. **Insect attractants** Some natural agents, as well as some synthetic substances can attract several insects. The instincts produced as a result of particular stimuli such as, light, smell, taste, touch etc. through receptors is well known. The smell coming out of food materials, the scent produced by the opposite sex, the odor from prey, the smell emanating from places ideal for egg-laying, phototropism and such other stimuli cause responses in insects and attract them towards the sources of such stimuli. These stimuli enable the insects to locate suitable mate for mating, locate food sources from long distances, locate suitable places for egg laying, and pupation and in various other life activities. Such insect behavior is taken advantage of in practice to attract insects to assess pest population in surveillance programmes and also to attract and destroy them.

   One such attractant is **'light'**. Many insect species are attracted to light, irrespective of the sexes. An electric lamp of 25 watt light intensity can attract insects from about 2.0 acres, whereas a petromax light with about 200 watt light intensity attracts insects from 10 - 15 acres. A petromax light can attract 20 times more of red hairy caterpillar moths than a hurricane lamp with an intensity of 15 - 20 candle power.

   Apart from light intensity, the color of the light spectrum also has varied effect on the insects. Except for a very few insects, most of the insects are much more strongly attracted by the blue portion of the light spectrum and by Ultra violet light than by the red and yellow portion of the light spectrum. This behavior of insects is taken advantage of in setting up **'light traps'** to attract and destroy many insect pests.

   Light trap in its simple form consists of a light source, which is hung over a wide - mouthed tray with water and some kerosene oil or crude oil on the water surface. Many types of light traps have been designed, of which the **'Robinson light trap'** is the most popular and widely used one. It consists of a wide - mouthed funnel with 3 metal baffles attached perpendicularly to the inner sloping surface of the funnel at equal distances. At the tail end of the funnel, a container is attached to collect the insects. The whole set-up is mounted on a tripod stand. A mercury vapor lamp is set up just above the mouth of the funnel. The insects, which are attracted by the light, fly around the light source, strike against the upright baffles in the funnel and fall into the container. The light trap is placed at a height of 2 - 3 metes above the ground level to get good results. It is enough to operate the light trap from 7.00 - 11.00 PM every day **(Fig. 9)**.

**Fig. 9** - Robinson light trap

Recently **'Black light traps'**, **'Solar light traps'** etc. have been introduced to monitor pest occurrence and for the control of crop pests under field conditions. In black light traps, the visible light is very poor, but it can attract insects from long distances. The black light trap is used for collecting many insects that are active and flying at nights and are attracted to ultra violet light, which is the light source in these traps. The insects attracted include many moths, as well as many other insects.

In solar light traps, the solar panels convert solar energy into electrical energy for illuminating the light source in the traps.

## Advantages of using light traps

Light traps help to catch all flying nymphs and adults of insects like leaf folder moths, stem borer moths, fruit borer moths, hoppers, fruit weevils and beetles and several other positive phototrophic insects.

i) The traps are portable across the crop area quite easily.

ii) They can be easily mounted and installed.

iii) They are very economical and helps in reducing the cost of pesticides and their application

iv) They are eco friendly device

v) They are durable and can be used year after year.

A few chemical compounds are used to attract and destroy some of the important crop pests. Some of these chemicals can attract either the male or female insects of a species, while some others attract both male and female insects of a species without any exception. **'Methyl eugenol'** attracts only the males of mango fruit fly - *Bactrocera (Dacus) dorsalis* even from a distance of about one kilometer. **'Anisyl acetone'** attracts the males of cucurbit fruit fly - *Bactrocera (Dacus) cucurbitae.* **'Fermented sugars'** and **'Molasses'** can attract several species of moths and butterflies. **'Ammonia'** attracts a few species of flies. **'Protein hydrolysate'** attracts many fruit flies. The synthetic chemical compound **'Siglure'** attracts the Mediterranean fruit fly - *Ceratitis capitata.* **'Metaldehyde'** attracts snails and slugs. Based on this fact, several poison baits have been formulated with a bait to attract the pest and a poison, mostly an insecticide to kill the pest.

Some of the commonly used poison baits are given below :

### Poison bait for *Spodoptera litura* caterpillars

| | |
|---|---|
| Rice bran (Preferably raw rice bran) | 5.0 kg |
| Molasses or jaggery syrup | 1.0 kg |
| Carbaryl 50 WP | 0.500 g |
| Water | Required quantity |

Rice bran, molasses or jaggery syrup and the insecticide are mixed with sufficient water to make a thick dough and made into small balls. These balls are placed along the field bunds and inside the fields in several places during the evening hours. The caterpillars, which come out from their hiding places for feeding after dusk are attracted by the smell of the bait, feed on the poison bait and are killed. The uneaten bait material should be collected the next morning and destroyed by burying, otherwise it will endanger the lives of cattle and other livestock that may eat it.

### Poison bait for sorghum and pearl millet shoot flies

Tin containers of 500 ml. capacity are taken and several holes of 2.0 mm. diameter are made on the side of the container from the outside. Dry fish soaked in water is put at the rate of 25 g. inside each container and a small quantity of Malahion 50 EC. is also poured inside and the container closed with the lid. These containers are placed in the field in several places at random. The flies, which are attracted by the smell of dry fish enter the container through the holes that are apparently smooth. Once they enter inside, they cannot come out because of the sharp edges on the inside of the holes and are trapped inside and eventually killed by the insecticide. The dry fish inside the container should be changed every week with fresh stock of fish.

### Poison bait for mango and cucumber fruit flies

| | |
|---|---|
| Methyl eugenol or Protein hydrolysate (Bait) | 1.0 lit. |
| Malathion 50 EC or Trichlorphon (Poison) | 0.1 lit. |
| Water | 10 lit. |

The bait and the poison are mixed with water, poured in small, open tin containers and are hung in the garden in several places. The flies, which are attracted by the smell of the bait, fall into the insecticidal solution and are killed.

### Poison bait for fruit - sucking moths

| | |
|---|---|
| Gur (Bait) | 1.0 kg |
| Vinegar (Bait) | 60 ml. |
| Malathion (Poison) | 50 ml. |
| Water | 10 litres |

The poison is mixed with the baits and kept in wide - mouthed vessels or pots in the orchard. The moths, which are attracted by the smell of vinegar and gur, fall into the insecticidal solution and are killed.

### Poison bait for rhinoceros beetle

| | |
|---|---|
| Castor seed cake | 1,0 kg. |
| Water | 5.0 lit. |

Castor seed cake is soaked in water and the ferment is poured in wide - mouthed pots and placed in coconut gardens. The beetles are attracted by the smell of the ferment, fall into the ferment and die. A small quantity of fresh cow dung may also be added, as the beetles are also attracted by the smell of fresh cow dung. The ferment should be changed every month.

Coconut toddy is another bait that can be used to attract and kill both rhinoceros beetles and red palm weevils. Coconut toddy is poured in wide - mouthed mud pots and placed in coconut gardens. Rhinoceros beetles and red palm weevils are attracted by the smell of toddy, fall into the toddy and die. The toddy should be replaced every 2 - 3 days.

### Poison bait for rats

| | |
|---|---|
| Zinc phosphide | 10 g. |
| Fried rice or sorghum in edible oil | 490 g. |

The poison and feed are mixed thoroughly and placed in places where rats frequent. The rats feed on the poison bait and die quickly. Pre - baiting with the food material alone for 2 - 3 days is necessary to make the rats eat the poison bait.

Ready made poison baits with Bromodiolone as poison and many other food materials and edible oil as bait is available in the form of cakes under different trade names viz., Robon, Rat kill etc. This poison kills the rats rather slowly by causing internal bleeding.

**7. Pheromone traps**

A **'Pheromone'** also known as an **'Ectohormone'** is a substance secreted by an insect to the exterior causing a specific reaction in the receiving insect. The behavior that is instigated in the receiving insect may be alarm, trail making, sexual activity, aggregation for mating, feeding or oviposition etc. This phenomenon of release of sex pheromone by one sex of a species to attract the opposite sex of the same species for mating is exploited in the use of pheromone traps to attract and destroy certain insects.

Several synthetic sex pheromones called **'Sex lures'** are used in pheromone traps to attract and destroy the male insects. The male insects, which are attracted by the sex lure are trapped inside the pheromone trap and are destroyed by means of an insecticide or immobilized by a coating of a sticky material and are killed ultimately.

**'Gossyplure'** a sex lure is used to attract the male moths of the cotton pink boll worm - *Pectinophora gossypiella.* **'Pherodine SL'** is a sex lure for the moths of the American cotton boll worm - *Spodoptera litura.* **'Disparlure'** is a sex lure for the gypsy moth - *Portheria dispar.* **'Bombycol'** is a sex lure for the silk worm moth - *Bombyx mori.*

**8. Repellents**

Some substances cause insects to move away from such sources and are called **'insect repellents'**. These are substances whose stimuli induce avoiding reactions in insects. Repellents are mainly of two types. One is the morphological characters of the host plants such as, presence of thorns, dense hairs, waxy coating or thick and hardy epidermis on the surface of the plant parts. The other is chemical repellents. Chemical repellents are widely used to protect man, cattle and plants.

**'Oil of Citronella'** and **'Camphor'** are widely used as mosquito repellents. **'Creosote'** is used to protect building timbers and other forest products, which are in contact with the ground from termite attack. Buildings are protected from termite attack by the use of **'Trichlorobenzene'**. Painting the stem portions of trees from the ground level with **'coal tar'** or **'crude oil'** also prevents termite infestation. **'Bordeaux mixture'**, which is primarily a fungicide, acts as a repellent to many biting larvae, grasshoppers etc. **'Paradichlorobenzene'** and **'Naphthalene'** moth balls repel insects, which infest woolen materials and other clothing. Chemical compounds such as, **Dimethyl phthalate'**, **'Diethyl toluamide' (Deet)**, **'Delphine'** etc. when applied to the skin protect the wearer against mosquitoes, ticks, mites, fleas and biting flies. **'Ant tapes'** containing **'Bichloride of mercury'** repel ants. **'Pyrethrum'** repels blood - sucking insects and is used as a spray on cattle in very small concentrations. Another repellent suitable for application on livestock is **'Butoxy polypropylene glycol'**. **'Synthetic pyrethroids'** are used in mosquito coils and mosquito mats to drive away mosquitoes.

A mixture of **'Sulphur powder'** and **'linseed oil'** repels rabbits, squirrels etc. **'Neem seed extract'**, **'neem oil'** and **'neem seed cake'** also have insect repellent properties. **'Copper sulphate'**, **'Tetramethyl thiuram disulphide'**, **'Brestan'** and **'Brestanol'** are some of the insect repellents by virtue of their bad taste to insects. Application of **'Pine tar oil'** or **'Diphenilamine'** repels screw worm flies from laying eggs around wounds of animals.

**'Sound'** at very high frequency also repels insects such as, mosquitoes, fleas etc.

9. **Antifeedants** or **Feeding deterrents**

Antifeedants are chemicals, which inhibit feeding of insects, thereby they are starved to death. Antifeedants are not repellents, as the insects are not driven or kept away from the food source. They act as inhibitors of gustatory reflexes and thus function only as feeding deterrents. The advantage of using antifeedants is their selective action against certain pests, but not against parasitoids, predators and insect pollinators. Antifeedants have their own limitations in that only surface feeding insects are prevented from feeding the treated plants, while the internal feeders and sucking pests are not affected. \ **'Triazenes'** have antifeedant property against several larvae, beetles and weevils, cockroaches etc. Some of the **'Organo tin compounds'** viz., Triphenyl tin acetate (Brestan), Triphenyl tin hydroxide (Du-ter) and Triphenyl tin chloride (Brestanol), which are actually fungicides, act as antifeedants against the larvae of potato tuber moth - *Phthorimaea operculella.* Some of the **'carbamates'** prevent feeding of some insect pests. The carbamate **'propoxur'**

**(Baygon)**, which has systemic action, acts as an antifeedant against the cotton boll weevil - *Anthonomus grandis.* **'Botanical extracts'** of a large number of plants possess antifeedant properties. **'Pyrethrum'** has antifeedant properties and act as feeding deterrents of blood - sucking flies and other non - insect pests. **'Neem products'** also have antifeedant properties against several insects. Several other compounds such as, plant growth hormones and chemicals are also known to develop antifeedant response in insects. The **'Plant growth hormones'** such as, **'Glycocel'**, **'Phosfan'**, **'Carvadan'** etc. act as feeding deterrents of cotton leaf caterpillars. Chemicals such as, **'Copper stearate'**, **'Copper resinate'** and **'Mercuric chloride'** have antifeedant properties.

**10. Electromagnetic energy**

Various forms of electromagnetic energy are made use of in the field of pest control. **'Radio frequency'** portion of the electromagnetic spectrum is effective in controlling many pests, especially those infesting grains, foodstuff and wood. Rice weevil infesting wheat grains is controlled effectively when exposed to radio frequency treatment for a few seconds.

**'Infrared radiation'**, which is a constituent of the electromagnetic spectrum is also used for the control of certain insects, especially stored grain pests. The heat energy produced by direct application of infrared radiation on infested stored material has been found to be quite promising in the control of certain insect pests. While rice weevil attacking wheat is more susceptible to infrared radiation, the Angoumois grain moth and the lesser grain moth have been found to be more resistant.

**'Ultraviolet radiation'** of the electromagnetic spectrum is also used to attract and kill certain species of insects. Exposure of insects to ultraviolet radiation damages the nucleic acids (DNA and RNA) in the insect body as a result the insects die.

**ix) Integrated control**

Several pesticides introduced in the recent past, though they are highly toxic to pests and can kill more than one pest very rapidly, there are certain disadvantages in using them. When some insecticides are used repeatedly, there is subsequent increase in the population of some pests. Two reasons are attributed to this phenomenon. One reason is, when some insecticides are applied repeatedly, some of the insect pests are not affected to a large extent, while their natural enemies are affected to a larger extent and are destroyed, thus the balance of nature is disturbed. As a result of this, the pests, which have not been destroyed by the insecticides multiply very rapidly and become more serious. This is known as **'pest upset'**. The other reason is, when the pests are kept in check by their natural enemies and the balance of life is maintained, if some insecticides are applied, the natural enemies of the pests are destroyed to a large extent and in the absence of their natural enemies, some of the pests multiply more rapidly and assume very serious proportions. This is known as **'pest resurgence'**. Under these circumstances, more and more highly toxic insecticides have to be used repeatedly to control the pests. This may besides polluting the atmosphere, cause health hazards, at the same time may result in wastage of time, labour and expenditure. This may also result in the development of resistance in the insect pests against certain insecticides and such resistance may even pass on to the ensuing generations. Further, due to the residual toxicity of highly toxic pesticides, harmful after effects may also result. To avoid such undesirable and dangerous trend, all pest control measures such as, cultural, physical, mechanical and biological methods have to be integrated along with chemical control methods to control the pests more effectively and to keep the pest level not to exceed the economic threshold level. This integrated approach is known as **'pest management'** or **'integrated pest control'**. Integrated pest control may vary from pest to pest and from place to place.

### Integrated pest management for rice

Immediately after harvest of the rice crop the field is ploughed so as to bury the stubble under the soil or removed from the field to prevent survival of the pests in the stubble or continued multiplication of the pests in the ratoon crop.

Alternate and collateral hosts of pests should be removed from the field bunds, irrigation channels and surrounding areas to prevent pests from continuing their lives and multiplying on these hosts in the absence of the primary host. The rice gall flies live and multiply in species of *Panicum, Paspalum* and *Cesania.* The rice mealy bugs continue their lives in species of *Cynodan, Cyperus, Eleusine, Eragrotis, Andropogan, Panicum, Ischaemum* etc. The rice leaf hoppers live and multiply in species of *Cynodan, Cyperus, Echinochloa, Eleusine, Eragrotis, Setaria* etc. throughout the year. Brown plant hoppers can live on species of *Leersia, Cesania* etc.

Trimming the bunds and pasting the sides with clay helps to bring out and destroy eggs of grasshoppers, at the same time destroy the weed hosts, which may harbor several pests.

During the summer months the field bunds and burrows in the fields are dug out and rats caught and killed.

Pest resistant or pest tolerant varieties may be cultivated. TKM.6 is resistant to stem borer and IR.20 has got certain amount of resistance to stem borer.

The seeds are cleaned thoroughly of chaff, damaged grains etc., dried well in sun and used for sowing. Otherwise the seeds may be put in 20 per cent common salt solution and stirred well. The chaff, damaged grains, pest infested grains etc., which float on the salt solution are removed and good seeds, which settle down at the bottom are washed thoroughly in several changes of water till they are completely free of the salt, shade dried and then used for sowing.

In places where white tip nematodes are prevalent, the seeds are treated in hot water at 52° - 55°C for 15 minutes, shade dried and used for sowing so as to destroy all stages of nematodes present inside the seeds.

Light traps may be set up from the very beginning of the crop to monitor the appearance of stem borer, leaf folder, leaf hoppers etc. and to destroy them.

The nurseries and transplanted fields are regularly monitored and egg masses of pests, deadhearts, larvae inside leaf folds, leaves with hispa larvae and other pests may be removed and destroyed.

Before planting, the seedlings are dipped in a solution of monocrotophos 36 SL - 0.04% to prevent infestation of stem borer, gallflies, leaf hoppers, brown plant hoppers etc. for up to 30 days after planting. This treatment protects the natural enemies of the pests.

In the transplanted crop, carbofuran 3G - 12 kg. or chlorpyriphos 10G - 4.0 kg. may be applied 30 days after planting to prevent stem borer, leaf folder and gall fly infestation.

When pests such as, leaf hoppers, brown plant hoppers, grasshoppers, army worms, earhead bugs etc. appear, suitable insecticides are applied at the correct dose to control them. While selecting insecticides, chemicals, which are less harmful to natural enemies of the pests may be selected. Systemic insecticides are less harmful to natural enemies than most of the contact poisons. Plant protection chemicals should be used only after the pest infestation crosses the economic threshold level.

## Integrated pest management for cotton

Acid delinted and fungicide treated seeds should be used for sowing.

Cultivation of alternate hosts of cotton pests such as, lady's finger, holly hock, tomato, egg plant, sunflower etc. near about cotton crop should be avoided.

Egg masses of pests, young caterpillars aggregating together on the under surface of leaves and grown - up larvae on the foliage may be collected and destroyed.

Infested leaves, plant parts, flowers, squares and bolls, which have fallen on the ground should be periodically collected and destroyed.

Application of excess quantities of nitrogenous fertilizers should be avoided and judicious water management practices should be followed.

Crops such as, sorghum, pearl millet, finger millet, groundnut etc. may be included in the crop rotation sequence with cotton.

Light traps should be set up at the rate of one trap per acre to monitor and destroy positive phototrophic insects.

Suitable pheromone traps should be set up at the rate of 5 traps per acre to attract and destroy the male moths of pink boll worm, *Prodenia* and *Heliothis.*

The appearance of various pests should be monitored from the very beginning of the crop, so as to take up control measures at the appropriate time.

Yellow sticky traps or Delta sticky traps should be set up at the rate of 10 traps per acre at a height of one foot above ground level in different places at random to destroy low - flying insects such as, aphids, white flies etc.

Only when the pest occurrence crosses the economic threshold level, suitable pesticides should be applied at the correct dosage for controlling the pests.

Repeated application of the same insecticide should be avoided. Suitable alternative chemicals should be applied every time, if necessary.

Repeated application of synthetic pyrethroids should be avoided, as they may induce resurgence in some of the sucking pests. Under unavoidable circumstances, they may be used once or twice to control the borer pests.

Synthetic pyrethroids should be followed by the application of an organophosphorus insecticide, if necessary.

Pesticides should be applied at the appropriate time when the pests are at their most vulnerable growth stage for their effective control.

Immediately after completion of harvest , the plants should be removed and disposed off to prevent perpetuation of the pests.

Ratooning of cotton crop should be avoided.

All efforts should be taken to maintain and protect the natural enemies of pests such as, parasitoids and predators, which occur naturally

The egg parasitoid - *Trichogramma evanescens*, the larval parasitoid - *Chelonus narayani* and the pupal parasitoid - *Chelonus rufus* may be released at the rate of 18,000 insects (1.0 cc.) per acre 3 - 5 times at 15 days intervals to destroy spotted boll worm and pink boll worm pests. The egg parasitoid - *Trichogramma chilonis* may be released to destroy American boll worm.

The predators - *Chrysopa cymbela* and *Chrysoperla cornea* may be released at the rate of 40,000 insects per acre to destroy white flies.\\

Poison bait may be used to attract and kill tobacco caterpillar. Pieces of blue cloth may be spread in-between plants in a few places during the evening hours to attract tobacco caterpillars, which hide under them. They can be collected and destroyed the next morning.

*Nuclear polyhedrosis virus* at 180 larval equivalent per acre along with cotton seed powder - 200 g., jaggery - 1.0 kg. and malathion - 400 ml. in 400 litres of water may be sprayed to control American boll worm.

*Nuclear polyhedrosis virus* at 100 larval equivalent per acre, along with protectants such as, crude sugar - 1.0 kg. in 400 litres of water may be sprayed to control tobacco caterpillar.

Only when absolutely necessary, pesticides should be used under proper guidance and supervision.

**Integrated pest management for groundnut**

Immediately after the receipt of summer showers, the fields should be ploughed to bring out the pupae of red hairy caterpillar to the soil surface and destroy them Sowing should be taken up in a synchronized manner.

Semi-spreading and spreading varieties are less susceptible to leaf folder attack than the bunch varieties. So, in leaf folder endemic areas, where protective irrigation facilities are available cultivation of semi - spreading and spreading varieties may be encouraged.

Proper water management practices should be followed to prevent the crop from drought, as drought conditions predispose groundnut crop to leaf folder attack.

Egg masses of pests, especially those of red hairy caterpillar , which are clearly visible to the naked eye may be collected and destroyed.

Castor or red gram crop may be raised as border crop or intercrop to attract red hairy caterpillars, which may be collected and destroyed easily.

Grown - up caterpillars can be hand picked and destroyed.

Light traps may be set up to attract and destroy the moths of red hairy caterpillar pest.

*Nuclear polyhedrosis virus* at 100 larval equivalent (100LE) in 300 litres of water may be sprayed to destroy young red hairy caterpillars.

Suitable pesticides should be applied only when absolutely necessary and when the pest incidence crosses the economic threshold level.

## Integrated pest management for pulses

The seed rate may be increased by 25 per cent and pest infested seedlings may be thinned out and destroyed to prevent loss due to stem fly infestation.

Sowing may be taken up early in the season to reduce stem fly and *Heliothis* infestation.

Blister beetles, grown - up hairy caterpillars, *Spodoptera* caterpillars and *Heliothis* caterpillars may be collected and destroyed.

Light traps may be set up to attract and destroy the positive phototrophic moths of *Spodoptera* and *Heliothis.*

Yellow sticky traps may be set up to destroy white flies.

Suitable *Nuclear polyhedrosis virus* may be used to control young caterpillars of *Spodoptera* and *Heliothis.*

Neem products such as, neem seed kernel extract - 5% may be used to control borer pests and white flies.

Appropriate pesticides may be used to control pests when absolutely necessary under proper guidance and supervision.

## Integrated pest management for sugarcane

Planting should be taken up early in the season, which helps in reducing shoot borer infestation to a considerable extent. In areas where stalk borer is very high, autumn planting should be avoided.

Soon after germination of sugarcane setts, a thick blanket of trash to a height of about 6 inches should be spread. This helps in suppression of shoot borer infestation. However, this practice may encourage termite, army worm and rat damage.

Application of excessive dose of nitrogenous fertilizers should be avoided in internode borer and *Pyrilla* endemic areas.

A light earthing up to cover the basal portion of the seedlings should be given 45 days after planting to prevent shoot borer infestation.

Dried and senescent lower leaves should be detrashed during the $5^{th}$ and $9^{th}$ months of crop growth and destroyed. This is effective in reducing infestation by several pests such as, internode borer, stalk borer, scale insects, mealy bugs, white flies and leaf hoppers.

Adult white grub beetles should be collected from their host plants at dusk and destroyed. This prevents white grub infestation of sugarcane plants to a large extent Pest infested canes may be removed and destroyed.

After completion of harvest, the stubble may be uprooted and destroyed so as to eliminate the pests harboring in the stubble or in the ratoon crop that may follow.

Sugarcane varieties having resistance or less susceptible to specific pests should be grown in such pest endemic areas.

Parasitoids and predators, which exist in the sugarcane ecosystem naturally should be encouraged. Wherever possible, suitable parasitoids and predators may be released. For the control of early shoot borer, *Granular virus* at $10^6$ - $10^7$ inclusion bodies per ml. along with a surfactant such as, Teepol or Sandovit - 0.05% may be applied on the 30$^{th}$ day after planting. This can be repeated at 15 days intervals, if necessary. In addition to *Granular virus,* 50 gravid females of the Tachinid larval parasitoid, *Sturmiopsis inferens* may be released per acre from the 30$^{th}$ to 50$^{th}$ day of planting. Further, the egg parasitoid, *Trichogramma chilonis* at 20,000 adults per acre may be released at weekly intervals. For the control of sugarcane top borer, 50 females of the pupal parasitoid, *Isotima javensis* may be released per acre when there is more than 10% top borer infestation. To control internode borer, a total of 1,00,000 egg parasitoids, *Trichogramma chilonis* per acre may be released at the rate of 10,000 parasitoids per acre during the 4$^{th}$ and 9$^{th}$ months and 20,000 parasitoids per acre during the 5$^{th}$, 6$^{th}$, 7$^{th}$ and 8$^{th}$ months after planting.

Suitable pesticides may be applied to control the pests if absolutely necessary.

# 3

# Pesticides

The chemicals, which can destroy pests or prevent their occurrence by their chemical action are known as **'Pesticides'**. Pesticides include **'insecticides', acaricides', 'nematicides', rodenticides', molluscicides'** etc., which are used to destroy insect pests, mites, nematodes, rodents, slugs and snails respectively.

Pesticides have been used even before the time when man never knew how to read and write. Even before 200 BC, several substances were known to control pests. Sulfur, arsenic poisons, fluorine poisons, as well as some plant extracts and oils were used to control pests. However, these substances could destroy only a few specific pests only to some extent.

## Advances in the invention of pesticides

The invention of many of the present day pesticides started only after 1867. **'Paris green'**, a synthetic chemical was the first pesticide used to control the **'Colorodo beetles'** of potato. Up to 1939, only a few inorganic chemical compounds and plant poisons were used as pesticides. Only after the Second World War much advances were made in chemical research. After the invention of DDT in 1939, a revolution had started in the field of pest control. Following the introduction of DDT, several other pesticides were identified and introduced for pest control. In 1941, BHC was invented by the British and French scientists. During the same period, organophosphorus pesticides were invented by the German scientists and parathion, malathion, demeton etc. were introduced.

Gradually more and more advances were made in the formulation of pesticides. Side by side many types of advanced plant protection equipments were also devised for quick and uniform application of pesticides. The newly introduced pesticides were found to be highly toxic and could kill many types of insects very quickly and had long residual toxicity and could have continued effect on the pests for much longer periods.

After the introduction of organophosphorus insecticides in 1944, systemic pesticides were invented. These chemicals are absorbed by the plant parts such as, seeds, roots, stems, leaves etc. and are translocated to all parts of the plants through the sap and can destroy sucking pests more effectively. The systemic pesticides such as, demeton, dimethoate, dimecron, phorate etc. were in use on a large scale till recently

In 1949, the carbamate pesticides were invented and some of the carbamates such as, carbaryl, propoxur, aldecarb, carbofuran etc. were in use till recently on a wide scale.

During the recent past, synthetic pyrethroids were synthesized and some of them viz., decamethrin, cypermethrin, permethrin, fenvalerate etc. have been found to be very effective, especially against borer pests.

### Insecticide formulations

Because of their high toxicity, most of the pesticides are not formulated and marketed in full strength. To facilitate easy handling, to reduce the risk due to high toxicity in the commercial formulations and to prolong their shelf life, the active material is dissolved in suitable solvents and mixed with stickers and spreaders and supplied in the form, which can be easily mixed in water and sprayed. Only a few pesticides are formulated with 100 per cent active material. These formulations, when mixed with specific quantity of water and sprayed, the active ingredient is distributed uniformly on the sprayed surface and destroy the pests, which come in contact with the chemical or destroy the pests, which ingest the chemical by feeding on the sprayed parts. Thus, each pesticide is formulated with a definite quantity of active ingredient along with other carrier materials in the forms, which can be sprayed, dusted or used as fumigants or applied in the form of granules.

**Dusts (D)**. In the dust formulations, the finely powdered toxic ingredients is mixed with well powdered saw dust, wood bark dust, sulfur, chalk, talcum, kaolinite clay, volcanic ash or any other such inert material and are used directly for dusting on the plant parts. The dusts generally have low concentrations of the active ingredient, which may vary from 0.65% - 25.0%. The efficacy of dusts depends on the particle size, and the property of the diluents. Mostly the particles of dusts are of size 100 microns or less. An ideal dust formulation should flow freely and should not cake or ball while in storage or when it is being used. The dusts have to be used with any type of dusters and dusting has to be done in the morning hours when there is dew deposition on the surface of the foliage and when the weather is calm with less of wind.

**Wettable powders (WP)** or **Water dispersible powder (WDP)**. The wettable powders are formulated in such a way that they can be mixed with water and sprayed. The active material is dispersed in water in the form of minute particles and held in suspension in water, but do not completely dissolve in water. They may contain more of the active ingredient, usually 15% - 95%. Along with the active ingredient some synergists or activators, stickers and spreaders are also added. The active material sticks on to the surface of the sprayed parts after the water evaporates, but do not enter into the plant tissues and so they are liable to be washed off easily. Because of the possibility of more uniform distribution of the active material on the sprayed surface, wettable powders are more effective than dust formulations. Wettable powders should have long shelf life, should disperse in water easily and uniformly, should be able to stick and spread on the sprayed surface without rolling down and should not form hard cakes in storage. In the case of wettable powders, the particles in suspension in the spray fluid may settle down gradually and hence a certain amount of agitation is necessary at the time of spraying to prevent the particles from settling down.

**Water -soluble powders (WSP** or **SP).** These are pesticides formulated in the form of dusts, but are readily soluble in water completely. Hence the active ingredient is not held in suspension as in the case of wettable powders, but is dissolved in water and so no agitation is necessary to prevent the active material from settling down. These formulations usually have high concentration of the active ingredient and therefore convenient to store and transport. Acephate 75 WSP, Cartap 50 WSP etc. are registered and marketed in India.

**Emulsifiable concentrates (EC)**. In these formulations, the active material is dissolved in solvents and emulsifiers and is available in a concentrated liquid form. When these formulations are mixed in water, the active materials do not dissolve in water, but are dispersed and held in suspension in water by the emulsifiers. When these formulations are mixed in required quantity of water and sprayed, first of all the solvents evaporate quickly, followed by the water and the active material alone sticks on to the sprayed surface. Because the particles are in a suspended state, there is a possibility of the suspended particles to settle down. So, a certain amount of agitation is necessary while spraying.

**Water -soluble concentrates (WSC)** or **Solution concentrates (SC)**. In these formulations, the active ingredient is dissolved in solvents only and no emulsifiers are added. When mixed with water for spraying, both the active ingredient and the solvents are completely dissolved in water. So, while spraying no agitation is necessary. When these are sprayed, the solvents and water evaporate leaving behind only the active ingredient on the sprayed surface. To improve the wetting property of the spray fluid, surfactants are added. Monocrotophos is a water - soluble concentrate.

**Insecticide solutions (S)**. Most of the synthetic organic pesticides are insoluble in water, but are soluble in solvents such as, Amyl acetate, Carbon tetrachloride, Ethylene dichloride, Kerosene oil, Xylene, Petroleum naphtha, Pine oil etc. Certain amount of active material is dissolved in specific solvent and sprayed directly without adding any water. Some of the solvents used in such insecticide solutions also have certain amount of insecticidal properties. Such insecticide solutions are mostly used for the control of household pests.

**Concentrate insecticide liquids**. In these formulations, the active ingredients in high concentrations are dissolved in non - vaporizing solvents. In some cases, some vaporizing solvents are also added so as to enable solution and droplet formation. These concentrate liquids have high density and are viscous in nature. They can be sprayed along with air under pressure as very fine particles in the form of mist over large areas with a very small quantity of concentrates. Usually no water is mixed with these concentrates before spraying. These concentrates can be sprayed from fairly good heights from aero planes or helicopters, and because they do not vaporize, settle down as mist on the crop surface. Malathion, Fenitrothion, Dimethoate, Phosphamidon, Toxophene etc. are available in the form of concentrate liquids for aerial spraying.

**Suspension concentrates** or **Flowables (F)**. In the case of some active ingredients, which are not soluble either in water or in solvents, they are formulated as flowables. The active ingredient micronized along with a solid carrier such as, inert clay and then dispersed in a small quantity of water. Prior to spraying, it is diluted with the required quantity of water. Sevin XLR containing carbaryl as active ingredient is a flowable formulation. It is safer to honeybees and has rain fastness.

**Insecticide aerosols**. In these formulations, the active material is dispersed into very minute particles of size varying from 0.1 to 50 microns in the form of smoke or mist that float in the air. This is made possible by burning the active material along with some other slow burning substance or heating the material gently by some heat source and turning it into a vapor state or the active material may be dispersed by some mechanical means into a vapor state. Sometimes some of the liquefiable gases are liquefied under high pressure and the active material is mixed with it. When the liquefied gas is let out through a minute hole, the active material comes out in the form of a mist along with the gas and is suspended in the air. This method is followed in the control of mosquitoes and other household pests. Aerosol containing **'propoxur' (Baygon)** as an active ingredient is widely used for the control of household insects.

**Fumigants.** These are sufficiently toxic chemical compounds, which vaporize under normal temperature and destroy insect pests. Fumigants are widely used for the control of storage pests in warehouses, godowns, merchant ships etc. They are also used to destroy soil - inhabiting insect pests, nematodes etc. Fumigants are used only under airtight conditions for good results. Most of the fumigants are in liquid state and may contain one or more poisonous fumes. The tablet made up of a mixture of Aluminium phosphide and Aluminium carbamate in proper proportions produces **'Phosphine'** or **'Hydrogen phosphide gas'** in the presence of moisture. This fumigant is available as **'Celphos - 3 g. tablets'**. When it comes in contact with moisture, chemical reaction takes place, as a result phosphine gas is liberated, which is highly poisonous to insect pests. **'Hydrogen cyanide'**, another highly poisonous fume is produced when calcium cyanide reacts with sulphuric acid slowly. In cases, where it is not possible to adopt spraying or dusting, fumigation can be adopted with success. Fumigants are also used for the control of rats, coconut red palm weevils etc. However, fumigation should be done with utmost care and under expert supervision, because of the highly hazardous nature. Some synthetic pyrethroids such as **'Prallethrin'**, **'Allethrin'** etc. are used in mosquito coils, mosquito mats and mosquito liquids for the control of mosquitoes.

**Granules (G)** or **Pelletted insecticides**. In granular formulations, the active ingredient is mixed with inert substances or carriers such as, lime, gypsum etc. and produced in the form of granules. When the carriers dissolve in water after application in the fields or leaf whorls, the active ingredients are released gradually in a controlled manner, get dissolved in the water and are absorbed by the plant parts and translocated throughout the plant tissues. Granules may be broadcasted in the field after stagnating some water in the field, so that the active material may dissolve in water and absorbed by the plants or they may be incorporated into the soil when there is sufficient moisture in the field or they may be placed in the leaf whorls as whorl application. Granules are highly suitable for spot application in particular parts of the field where there are pest infestation or around tree trunks in shallow trenches followed by irrigation. Granules are of varying sizes ranging from 250 - 1250 microns in diameter. Granules of size ranging from 100 - 300 microns are referred to as micro granules. Granules may contain 2% to 10% of the active ingredient.

The effectiveness of granular insecticides may vary depending upon the uniformity in application, rainfall, temperature, size of the granules, physical and chemical properties of the soil, and the nature of the pests. Granules containing systemic insecticides can control sucking insect pests effectively,

while those containing non - systemic contact and stomach poisons can control soil - inhabiting pests and biting pests. Whorl application of granular insecticides is effective in controlling stem borer pests of sugarcane, sorghum, maize etc. Different systemic formulations are used for the control of weeds, plant diseases, nematodes, snails and slugs and rodents besides insect pests.

Application of granular insecticides is very easy. Granules can be used in certain specific areas alone as spot application instead of application in the entire field. The chances of the granular insecticides to be blown away by wind are very much less. They may not stick on to the surface of plant parts and hence the toxic effect on the surface of the plant parts is very minimal. The active ingredient in granules is released slowly and so their residual toxicity persists for a much longer period in the plant parts. For applying granular insecticides, there is no need for water. They are ready to use and require no mixing. They are less hazardous to the applicator. Granular insecticides have no significant adverse effect on the natural enemies of pests.

However, there are some disadvantages in the use of granular insecticides. They are not as effective as spray application against crawling insect pests. They are more expensive than wettable powder and emulsifier concentrate formulations. They require moisture for activating pesticidal action.

Types of different Plant Protection chemical formulations are given in **Appendix - 10**.

**Poison baits**. In poison baits, a primary food material or an insect attractant called **'lure'** is mixed with a small quantity of pesticide as **'poison'**. Several substances such as, rice or wheat bran, molasses or fermenting sugars, vinegar, Amyl acetate, orange fruits, Methyl eugenol, Protein hydrolysate, dried fish, Formaldehyde, Ammonia etc. are used as lure to attract specific pests. Pesticides such as, Malathion, Lead arsenate, White arsenic, Zinc phosphide etc. are used as poison in various poison baits. Biting insect pests, soil - inhabiting white grubs, fruit sucking moths, shoot flies, fruit flies, household pests, slugs and snails, crabs, rats etc. can be effectively controlled by the use of poison baits. When other pest control methods such as, spraying, dusting etc. fail to afford effective control of certain pests, this method can be followed.

**Encapsulated formulations**. Some highly toxic pesticides are packed in gelatin capsules and used for pest control. When such formulations are applied to the soil, the gelatin cover dissolves in water first and then the systemic pesticides are released into the soil, which dissolves in water and are absorbed by the plant roots and translocated into all parts of the plants, thereby afford protection against pests. This method is also highly suitable for spot application

**Insecticide - fertilizer mixtures**. Sometimes pesticides are mixed with fertilizers and applied at the time of regular fertilizer application. The mixture supplies both plant nutrients, at the same time controls insect pests, especially soil - inhabiting pests. Granular formulations may be conveniently applied by this method without incurring additional expenditure for pesticide application.

Systemic granular formulations when applied by this method gives effective control of sucking insect pests. Spraying of 2% urea solution mixed with compatible wettable powder or liquid insecticide formulations, besides supplying nitrogen to the plants, controls specific pests simultaneously.

## Additives in pesticides

**Stickers**. These are substances added to pesticide formulations to make the sprayed particles to stick on to the sprayed surface and not rolling down. Some types of clay, Mineral and Vegetable oils, Casein, Gelatin, Soybean flour etc are used as stickers.

**Spreaders and wetting agents**. These are substances added to pesticide formulations to decrease the surface tension of the spray fluids, so that the spray droplets can spread freely and easily over larger areas of the sprayed parts. Soaps, Teepol, Tergitol, Triton, Mersolate, Saponine of plant origin, Castor oil, Linseed oil, Cotton seed oil etc. serve as spreaders.

**Synergists** or **Activators**. Certain compounds, though they are non - toxic or slightly toxic to insects by themselves, such as, Pyrethroids, Rotenoids, Nicotine etc., when added to certain Organophosphorus and Carbamate pesticides, increase the toxicity of the mixture much more than the toxicity of the individual pesticide. This phenomenon is known as **'synergism'**or **'activation'**. Piperine from black pepper, Sesamin from sesame oil, Thanite, Terpin diacetate, Piperonyl cyclonene etc. act as synergists.

**Stabilizing agents**. To retard decomposition of certain unstable organic insecticides in storage, such substances are added. Epichlorohydrin, an acid inhibitor prevents decomposition of Aldrin and Toxaphene formulations. Organochlorine insecticides are stored in metal drums provided with special interior lacquer coatings to prevent decomposition.

**Deodorants and masking agents**. These are substances added to certain pesticide formulations to mask the unpleasant odors of the pesticidal components such as, Pyrethrins, Thiocyanates and Methylated naphthalene. Pine oil, some Flower scents etc. are added to insecticidal formulations at 0,1% - 1.0% serve as masking agents.

## Classification of insecticides

Insecticides are classified in different ways based on their mode of entry into the body of insects, mode of action on the insects and on the chemical nature of the toxicants.

### 1. Classification based on the mode of entry

i) **Stomach poisons**. These poisons can act only if they are ingested by the pests. The pesticide may enter into the alimentary system of insects, when they feed on the pesticide sprayed or dusted plant parts. Once they enter into the alimentary system, they interfere with the functioning of the digestive system and destroy them. These types of pesticides are mostly meant for killing biting and chewing type of pests. For obtaining effective control with such poisons, they should not affect the palatability of the hosts and should be quick acting. They may be applied as dusts, sprays or as poison baits. Lead arsenate, Calcium arsenate, Paris green, Arsenic pentoxide, Sodium fluoride, Cryolite, Lime - sulphur, Zinc phosphide, Mercuric chloride, Boric acid etc. are stomach poisons. Many synthetic pesticides belonging to various groups are stomach poisons. Most of the systemic pesticides are also stomach poisons. Systemic poisons are absorbed by the various plant parts and are translocated throughout the plant tissues. When sucking pests feed on the plant sap, the pesticides also enter the stomach of the pests along with the plant sap and destroy them

**ii)** **Contact poisons**. These poisons can enter into the body of the pests either through the cuticle or the soft skin in-between the body segments, antennal sockets or the respiratory spiracles. The pests may come in contact with the pesticides directly while spraying or dusting or while moving on the pesticide applied plant parts. They affect the nervous system of the pests, inactivate the functioning of the organs and destroy them. Contact poisons are capable of killing sucking pests also. Most of the Organochlorine, Organophosphorus, Carbamae and Synthetic pyrethroids are contact poisons.

**iii)** **Fumigants**. These pesticides when applied turn into gaseous state, enter the body of the pests through the respiratory spiracles, affect the respiratory enzymes or functioning of the gaseous exchange system in the cells, thereby destroy the pests quickly. All kinds of insects can be controlled by these pesticides, irrespective of their feeding habits. Phosphine, Hydrogen cyanide, Methyl bromide, EDCT etc. are well known fumigants.

Some of the pesticides may enter the body of pests and act in different ways. Lindane, which is primarily a contact insecticide, acts as a stomach poison and fumigant. Several pesticides are both contact and stomach poisons.

## 2. Classification based on the mode of action

**i)** **Physical poisons.** (a) These poisons affect the body of the pests by exerting a physical effect and destroy them. Mineral oils, Coal tar and such other chemicals physically block the respiratory spiracles, obstructing the insects from breathing and bring about their death due to asphyxiation. Such poisons are highly effective in controlling the scale insects.

(b) Inert materials such as, Aluminium oxide are mixed with some dust formulations. These substances, because of their rough surface and sharp edges make small aberrations on the epicuticle of the pests, through which the body fluids are drained leading to their death . Silica gel also acts in a similar manner.

(c) Some of the inert materials such as, Volcanic ashes added to some dust formulations as carriers, are capable of absorbing the body fluids of the pests due to their hygroscopic nature and cause their death. Activated clay also acts in a similar manner and destroys pests, especially storage pests

ii) **Protoplasmic poisons.** These are poisons, when ingested affect the protoplasm of the epithelial cells lining the alimentary tract of the pests and transform the protein into a hard crystalline form, thereby affecting the digestion process and destroy them. Mercury, Copper and such other heavy metal compounds, some Fatty acids, Formaldehyde, Ethylene oxide, Nitrophenols, Fluorine compounds etc. have such mode of action.

iii) **Respiratory poisons.** These poisons act on certain enzymes of the cells and affect the gaseous exchange system in the cells of the pests and cause their death. Hydrogen cyanide, Aluminium phosphide, Methyl bromide, EDCT etc. act in this manner.

iv) **Nerve poisons.** These pesticides obstruct the production of certain enzymes such as, Acetyl cholinesterase in the cells, affect the nervous system, inactivate the functioning of the organs of the pests and destroy them very quickly. Most of the Chlorinated hydrocarbons, Organophosphorus compounds, Carbamates, Nicotine, Pyrethroids etc. are nerve poisons

v) **Poisons of a more general action.** Some poisons such as, Chlordane, Topxaphene, Aldrin, Dieldrin etc. induce nervous debility in the pests and destroy them. Ryania, Phenothiazene and similar poisons affect the muscular action of the pests and kill them.

### 3. Classification based on the chemical nature

Based on the chemical nature, pesticides are broadly classified into two groups viz., Inorganic compounds and Organic compounds.

| Pesticides | |
|---|---|
| **Inorganic compounds** | **Organic compounds** |
| Arsenic compounds | Mineral oils |
| Fluorine compounds | Pesticides of animal origin |
| Sulphur and Sulphur compounds | Pesticides of plant origin |
| Zinc compounds | Synthetic organic compounds |
| Other inorganic compounds | |

## Inorganic compounds

**Arsenic compounds**. Among the arsenic compounds, the arsenates are generally more stable and safer for application than the arsenites. The arsenites, because of their phytotoxic nature on plants, are not used on plants, but are mainly used in poison baits. On the other hand, the arsenates can be used directly on plants, because they are less toxic to plants than the arsenites. However, the arsenates are less toxic to insects than the arsenites.

i) **Calcium arsenate**. This insecticide is formulated as a wettable powder with 25 - 30 per cent of active material. The formulation at 0.3 - 0.6 kg. along with the same quantity of hydrated lime is mixed in 200 litres of water and used for spraying. If lime is not added to this formulation, it may cause injury to the plants as a result of phytotoxicity. It is a stomach poison and can control chewing insects more effectively. It was widely used for the control of eggplant spotted beetle pest till recently. However, with the introduction of several newer insecticides, which can be directly mixed with water and sprayed and which are capable of controlling many pests simultaneously, its use has been limited to a large extent. Further, the formulation has a short shelf life and if stored for more than one year, it undergoes chemical changes and its effectiveness is lost.

ii) **Lead arsenate**. It is primarily a stomach poison with some amount of contact action. It is applied either as a dust or spray. For dusting, it is mixed with 3 - 20 parts of talk, sulphur dust, gypsum or hydrated lime. For spraying, 1.5 - 2.0 kg. of lead arsenate is mixed with 500 litres of water and used for the control of gypsy moth, lemon butterfly, lady bird beetle, red pumpkin beetle and castor semilooper. It is also used in poison bait for the control of fruit sucking moths.

iii) **Paris green**. The main ingredients in Paris green are copper acetate and copper arsenate. It contains 33 - 39 per cent of metallic arsenic. It is applied either as dust or spray. Spray solution is prepared by mixing 250 - 300 g. of the chemical compound and 1.0 kg. of lime in 500 litres of water. For dusting Paris green is mixed with 4 - 5 times its volume of lime. It is used for the control of Colorado beetles of potato, cotton pests and mosquito larvae. It is also used in poison baits for the control of slugs and snails.

iv) **White arsenic** or **Arsenous oxide**. It consists of 75 per cent arsenic and is used in poison baits for the control of rats, grasshoppers, army worms, ants and cockroaches.

**Fluorine compounds**. Fluorine compounds have been used as insecticides since 1800. They are primarily stomach poisons with some contact action. They are more effective than arsenic compounds in effecting a more rapid kill of insects. Fluoride poisoning produces spasms, regurgitation, paralysis and ultimate death.

i) **Sodium fluoride**. It is used in poison baits for the control of cockroaches, earwigs, cutworms and grasshoppers. The powder is also used to control biting lice on livestock and poultry. It is highly phytotoxic.

ii) **Cryolite**. It is known as 'Sodium flualuminate'. It is used both as dust and as spray for the control of chewing insects. For dusting, 20 - 60 per cent dust mixed with carriers such as, clay, talc or lime is used to control flea beetles and several other chewing insect pests. For spraying, 1.5 - 4.0 kg. of the powder is mixed with 500 litres of water and then sprayed.

**Zinc compounds**

**Zinc phosphide**. Zinc phosphide in the form of 2.0 per cent poison bait is used against locusts, crickets, moles and rats. It is one of the most widely used poisons against rats. It is a gray - colored powder with the odor of garlic. The chemical when ingested along with the bait material reacts with the hydrochloric acid found in the intestine of rats and phosphene gas is liberated. This highly poisonous gas affects the nervous system of the rats and kills them quickly.

**Other inorganic compounds**. Several other inorganic chemical compounds possess insecticidal properties, while some others serve as attractants or repellants.

**Mercuric chloride**. Mercuric chloride was widely used against soil - inhabiting insect pests infesting radish, cauliflower, cabbage etc. However, its use has been almost completely restricted because of its very high toxicity. Mercuric chloride and mercurous chloride are used as repellants for cockroaches, termites and ants in book binding.

**Thallium sulphate**. Thallium sulphate is used against ants in poison baits, mixed with sugar or jaggery

**Borax** or **Sodium tetraborate**. Borax is used to destroy fly maggots found in manure pits and in the wounds of cattle. Boric acid is used as a stomach poison to destroy cockroaches.

**Formaldehyde**. Formaldehyde is used in poison baits against house flies.

**Bordeaux mixture**. Bordeaux mixture, which is a mixture of copper sulphate and lime, besides being a potent fungicide, acts as a repellant against flea beetles, grasshoppers and some other insects.

**Barium carbonate**. Barium carbonate is used against rats in poison baits.

## Organic compounds

**Petroleum oils** or **Mineral oils** or **Hydrocarbon oils**. The constituents of mineral oils are hydrogen and carbon. Mineral oils are derived from sedimentary rocks. Kerosene oil, lubricating oils, tar, asphalt and other such compounds are obtained from them. Among coal tar oils, creosote and anthracene are used against certain insect pests. Though mineral oils in their natural state are phytotoxic, they serve as important solvents or diluents for chemical toxicants.

**i) Kerosene oil - soap emulsion.** The following materials are required for preparing this emulsion

| | |
|---|---|
| Kerosene oil | 20 litres |
| Soap | 0.5 kg. |
| Water | 10 litres |

The water is boiled first and powdered soap is put into the boiling water. To this mixture kerosene oil is added and stirred vigorously till a emulsion is formed. To this cooled stock solution water is added in the ratio 1 ; 20, stirred well and then used for spraying. It is effective in controlling all kinds of sucking pests.

**ii) Crude oil emulsion.** The following materials are required to prepare this emulsion.

| | |
|---|---|
| Crude oil | 20 litres |
| Fish oil soap | 0.5 kg. |
| Water | 10 litres |

This emulsion is also prepared in the same way as kerosene oil - soap emulsion. To the stock solution water is added in the ratio 1 : 20 and used for spraying. It controls termites and all kinds of sucking pests. Sugarcane setts dipped in this insecticidal solution before sowing prevents termite infestation. Termites are also controlled in the field by mixing 1,000 litres of this insecticide with irrigation water and irrigate the field.

These insecticides are very cheap and can be prepared quite easily. They have good spreading quality. They are non - toxic to man and livestock. Insects do not develop resistance to such type of insecticides. However, they are comparatively less toxic to insect pests and may cause injury to plants. They corrode the spray equipments and may damage them. Further, they have very short shelf life and are rather cumbersome to prepare.

**Organic compounds of animal origin**

i) **Nereistoxin**. Poisons obtained from certain animals may have insecticidal properties. The most important one is 'Nereistoxin', which is obtained from some marine annelids. Synthetic compounds having the properties of Nereistoxin have been synthesized and are being used.

   One such chemical is **'cartap'**, which is available under the trade names 'Padan', 'Thiobel', 'Vegetox' etc. as 50 per cent wettable powder, 2.0 per cent dust and 10 per cent granules. It is a contact and nerve poison. It controls pests such as, rice stem borer, rice leaf folder, sugarcane shoot borer, cabbage diamond - back moth etc. effectively.

ii) **Fish oil - rosin soap**. The following materials are required for preparing this insecticide.

| | |
|---|---|
| Powdered rosin | 10 kg. |
| Caustic soda | 3.0 kg. |
| Fish oil | 1.5 kg. |
| Water | 500 litres |

   Rosin and fish oil are heated first in 10 litres of water. When the rosin has dissolved completely, caustic soda dissolved in some water separately is added to the mixture and stirred vigorously. Then water is added to make up the volume to 500 litres. This insecticide can control many sucking pests, especially thrips, scale insects, mango hoppers etc. It is commercially available as an emulsion and is mixed with water in the ratio 1 : 80 for spraying.

**Organic compounds of plant origin**. Several plant products are being used as insecticides, nematicides, attractants, repellants and diluents and some of them are widely used in the field of plant protection.

i) **Nicotine**. This plant poison is obtained from the leaves of tobacco plants, *Nicotiana tabacum, N. rustica* and some other species of *Nicotiana,* as well as from some other plant species such as, *Asclepias syriaca* etc. Of the twelve alkaloids present in tobacco, nicotine is the most important one and is found to the extent of 2,0 - 14.0 per cent. It can enter into the body of insects either through the cuticle, through the mouth or through the respiratory spiracles. It acts as a nerve poison and kills the insects very quickly. It has also got fumigant action. It is used extensively for the control of cardamom thrips. It is also effective in controlling aphids, mealy bugs etc.

Tobacco decoction is prepared from dry tobacco leaves as follows:

| Dry tobacco leaves | 10 kg. |
|---|---|
| Soap | 0.1 kg. |
| Water | 30 litres |

The dry tobacco leaves are boiled in 10 litres of water for 10 minutes or soaked in water for 24 hours. Then the whole thing is cooled and strained. Water is added to this decoction to make up the volume to 30 litres. Soap is then added to this decoction as shavings and the mixture is again heated till the soap dissolves completely and then cooled. This solution is used for spraying. Casein may be added to tobacco decoction as a spreader. The insecticide does not leave any toxic residues on the sprayed plant parts. Further, it does not pollute the atmosphere. However, its use has been very much limited because of the difficulties involved in its preparation and also because the prepared decoction cannot be stored for many days.

Nicotine sulphate, which is a bye - product of tobacco manufacture, is commercially available as an emulsifiable concentrate containing 40 per cent nicotine and as a dust with 2.0 per cent nicotine. The dust formulation can be dusted directly on the plant surface . The emulsifiable concentrate is mixed with water in the ratio 1 : 500 to 1 : 1,000 and soap (2.0 - 3.0 kg. of shavings per 500 litres) or lime is added and then sprayed. It controls thrips, aphids and other soft - bodied insects effectively.

ii) **Pyrethrum**. It is obtained from the flowers of certain species of pyrethrum plants such as, *Chrysanthemum cinerariaefolium C. cocineum, C. roseum, C. carneum* etc. The active ingredients of pyrethrum are the esters of 'pyrethrins' and 'cinerins', which are extracted from the flowers. The flowers are dried, coarsely powdered and subjected to solvent extraction to separate the active materials. Pyrethrum dust is prepared by drying and powdering the flowers and adding inert materials or carriers such as, talc, clay or lime. Pyrethrum is formulated as dust, emulsifiable concentrate, solution or aerosol and in most of the formulations the pyrethrum content ranges from 0.05 - 0.1 per cent. The emulsifiable concentrate used for spraying contains 2.0 per cent pyrethrins and the dust formulations contain 0.1 - 0.2 per cent pyrethrins. Powdered and extracts of pyrethrum are widely used against flies, bed bugs and silver fishes. Dust and spray formulations are used against insect pests infesting vegetables, ornamental plants and fruit trees. Pyrethrum is a powerful contact insecticide and has quick knock - down effect, but not very effective as a stomach poison. Sprays containing 0.002 - 0.004 per cent pyrethrins kill most insects. It is also used as a cattle spray for the control of ectoparasites. Pyrethrum leaves no toxic residues on the treated plant parts. Pyrethrum and the synergist, piperonyl butoxide are mixed together in the proportion 1 : 10 to obtain the insecticide **'Pyrocone'**, which can control coconut red palm weevil very effectively. Pyrethrum mixed with stored grains affords protection against stored grain pests.

**iii) Rotenone**. Rotenone or **'Rotenoid'** is a poison present in some species of leguminous plants such as, *Derris elliptica, D. malaecensis* etc., which contains 5.0 - 9.0 per cent rotenone and *Lonchocarpus utilis, L. uruca* etc., which contain 8.0 - 11.0 per cent rotenone. Rotenone is the main toxic constituent obtained from the roots of these plants. The roots are dried, powdered and mixed with 3 - 7 parts of carriers such as, talc, clay or gypsum etc. to be used as dust. It is also extracted from the powdered roots, which is used in spray formulations. It is used as dust containing 0.05 - 0.1 per cent rotenone or 1.75 - 3.50 per cent of total extractives for the control of pests infesting horticultural crops. Rotenone kills insect pests by attacking the nervous system, causing paralysis. It does not leave any toxic residues on the treated plant parts for more than 5 - 10 days, It is highly toxic to fishes and other water inhabiting organisms.

**iv) Ryania**. It is obtained from the roots and stems of the plants, *Ryania speciosa* and its most important toxic ingredient is the alkaloid, **'Ryanodyne'**. It is both a contact and stomach poison, but it is less toxic to mammals than rotenone. It is more stable and possess longer residual action. It is used for the control of Lepidopterous insect pests such as, codling moth, maize borer, sugarcane borers etc.

**v) Sabadilla**. It is an alkaloid poison obtained from the seeds of lily sabadilla viz., *Schoenocalon officinale.* The seeds contain 2.0 - 4.0 per cent of a crude mixture of the alkaloid called **'Vertrine'**. It is a contact and stomach poison and highly toxic to mammals. It is used for the control of human lice, house flies and other household insects as well as Hemipterous plant pests

**vi) Tephrosin**. Tephrosin is another plant poison obtained from a few species of *Tephrosia.* It is found in the leaves and seeds of *Tephrosia vogelii,* from the roots of *T. toxicaria* and from the stems of *T. macropoda*. It is primarily a contact poison and can control several insect pests.

vii) **Sweet flag**. This indigenous plant, *Acorus calamus* is commonly found in the Western Ghats, Nilgiris and a few other parts of the country. The rhizomes of sweet flag possess insecticidal properties. Mixing 1.0 kg, of the dried and powdered rhizomes with 50 kg. of grains affords protection from storage pests for about one year. Spraying an infusion prepared by soaking the powdered rhizome at 7.5 - 10.0 g. in 5.0 litres of water along with an equal quantity of soap controls aphids, leaf feeding caterpillars etc.

viii) **Thevetia**. Decoction prepared from the seeds of this garden plant, *Thevetia neriifolia* is found to be effective against both chewing and sucking pests. Seeds of this plant are dried and powdered and 15 - 30 g. of the powder is soaked in 10 litres of water along with the same quantity (15 - 30 g.) of soap. The filtered decoction is used for spraying. It controls several insect pests such as, aphids, white flies, thrips, scale insects, leaf feeding caterpillars etc.

ix) **Neem**. The neem tree, *Azadirachta indica* is a very important deciduous tree commonly found throughout India. Various parts of this tree possess gustatory repellant and insecticidal properties. Neem seed kernels have been reported to possess extraordinary gustatory repellant properties and the active ingredients responsible for such repellant action are **'Nmbidin'**, **'Azadirachtin'** and **'meliantriol'** and these toxic materials do not produce any harmful effect on the treated plant parts. Neem seed powder suspension at 0.1 per cent concentration (10 g. of seed powder in 10 litres of water), when sprayed, acts as a repellant to desert and migratory locusts. Powdered neem seed kernel, when mixed with wheat seed at 1 - 2 parts per 100 parts of seeds affords protection against storage pests such as, rice weevils, lesser grain borers and khapra beetles for about one year and the treatments has no adverse effect on the germination of the seeds. It also protects black gram, Bengal gram, cowpea and pea from pulse beetle infection for about one year. Neem seed cake extract is prepared by soaking 1.0 kg. of neem seed cake in 5.0 litres of water for 7 days with frequent stirring, then filtering the extract through a cloth and making the volume of the filtered extract to 10 litres by adding water. This extract is sprayed on citrus plants to minimize citrus leaf miner infestation. Neem oil emulsion is prepared by mixing neem oil - 4.0 litres and liquid soap - 200 ml. in 200 litres of water and stirring the mixture vigorously to obtain 2.0 per cent neem oil emulsion.

It is used to control cotton white flies, gingelly shoot webber, coconut mealy bugs, rice green leaf hopper and pests infesting garden plants. Neem oil - 3.0 per cent emulsion is used to control rice brown plant hopper. Neem seed kernel extract - 5.0 per cent is prepared by soaking 10 kg. of powdered neem seed kernel in 50 litres of water for 12 hours. Then the soaked powder is crushed and squeezed and the extract filtered through a cloth. To this extract 200 ml. of liquid soap is added and stirred vigorously. To this mixture water is added to make up the volume to 200 litres and is used for spraying. This extract can control pulses white flies, rice brown plant hopper and rice black bugs. Applying powdered neem seed cake - 10 kg. along with farm yard manure - 10 kg. per acre around the stem region reduces the infestation of cotton stem weevil.

**'Azadirachtin'** is extracted from the kernels of neem seed and is available commercially in the form of a powder, as well as in a emulsifiable concentrate form. Besides being an antifeedant, it disrupts insect moulting by antagonizing the hormone ecdysone

**'Nimbidin'** is another alkaloid obtained from neem product and is introduced for pest control in India.

**x)** ***Lobelia excelsa***. The leaves of this plant, which are found in the Nilgiris and Western Ghats have insecticidal properties. 1,0 kg. of leaves of these plants is cured under shade and chopped. The chopped leaves are soaked in water for 24 hours, filtered and the volume of the filtrate made up to 20 litres. To this solution 60 g. of soap is added and stirred vigorously and then used for spraying. It controls aphids on snake gourd and cowpea, tingids on eggplant and mites on castor and lady's finger.

**xi)** ***Pongamia glabra***. Pongam cake obtained from the seeds of pongam trees - *Pongamia glabra* has been found to possess insecticidal properties. Application of pongam cake to the soil controls ground beetles infesting newly transplanted tobacco seedlings.

**xii)** **Garlic**. Insecticidal property of garlic - *Allium sativum* is known for a long time. Garlic oil obtained from garlic bulbs contains **'Diallyl disulfide'** and **'Diallyl trisulfide'**, which have larvicidal properties and can be used for the control of mosquito larvae.

**xiii)** **Nematicides in plants**. Certain plant species have nematicidal or nematostatic properties and can suppress nematode population in the soil. Root exudates of the Crucifers viz., *Brassica nigra* and *Sinapis alba* inhibit the emergence of the golden cyst nematode - *Globodera rostochiensis* due to the presence of **'isothiocyanates'** in the root diffusates. Root extracts of *Asparagus racemosus* inhibit hatching of eggs of the root knot nematodes - *Meloidogyne javanica* and *M. arenaria.* **'Catechol'** present in the root diffusates of *Eragrostis curvula* suppresses *Meloidogyne.*

xiv) **Volatile oils from plants**. The volatile or ethereal oils are obtained from special glands of some plant species and their pungent odors are characteristic of the plant source. These oils are not greasy or viscous. They are chiefly used as attractants in baits or as repellants. Eugenol and Geraniol attract certain insect species. Eugenol is used widely in poison baits to attract and destroy fruit flies. Citronella oil, Cedar oil etc. are repellants and are widely used to drive away many insects. Menthol, Camphor, Peppermint oil and some such volatile oils also possess repellant action on many insects.

**Synthetic organic compounds or Synthetic organic pesticides**

**Organochlorine compounds or Chlorinated hydrocarbons**

i) **Dicofol ($C_{14}H_9Cl_5O$)**. Technically it is 'Trichloro bis (chlorophenyl) ethanol'. It is a contact and stomach, non-systemic, specific acaricide. It can destroy mites at all growth stages and it has got a quick knock down effect on mites. It is used to control mites on a wide range of crops including fruits, vines, ornamental plants, vegetables etc. It is widely used for the control of mites in tea. It is harmless to honeybees. Its residual toxicity lasts for a very long time. It is formulated as 18.5% EC and marketed under the trade names 'Difol', 'Kelthane', 'Delcofol', 'Hexakel', 'Decofol' etc. It is applied at the rate of 400 ml. per acre as spray.

**Organophosphorus compounds**. The first organophosphorus insecticide developed was **'Tetraethylpyro phosphate'** (TEPP) as a substitute for nicotine. This was followed by the development of Parathion and Shradan. **'Shradan'** was the first practical systemic insecticide, named after **Gerhard schrader**, who developed this insecticide along with his associates. After this several other organophosphorus insecticides have been developed and are being used as effective insecticides. Among these, there are contact and stomach poisons, systemic ones and some with fumigant action. Some of them are broad spectrum insecticides and are effective in controlling a wide range of insect pests, while some have acaricidal action also. However, most of these insecticides are relatively highly toxic to humans and mammals. Insecticidal activity of organophosphorus compounds is attributed to their ability to inhibit the enzyme cholinesterase in the insect nerve tissue, thereby disrupting the normal functioning of the nervous system, leading to their death. Poisoning in insects by organophosphorus insecticides causes hyperactivity, tremors, convulsions, paralysis and ultimate death. Among these, there are contact and stomach poisons, systemic ones and some with fumigant action. Some of them are broad spectrum insecticides and are effective in controlling a wide range of insect pests, while some have acaricidal action also. However, most of these insecticides are relatively highly toxic to humans and mammals. Insecticidal activity of organophosphorus compounds is attributed to their ability to inhibit the enzyme cholinesterase in the insect nerve tissue, thereby disrupting the normal functioning of the nervous system, leading to their death. Poisoning in insects by organophosphorus insecticides causes hyperactivity, tremors, convulsions, paralysis and ultimate death.

## Organo phosphorus compounds (Non - systemic)

i) **Chlorpyriphos ($C_9H_{11}Cl_3NO_3PS$).** Technically it is 'O, O - diethyl O - (3, 5, 6 - trichloropyridyl - 2) phosphorothioate'. It is a chlorinated organophosphate, broad spectrum, contact and stomach poison and can control various sucking and chewing insect pests and mites attacking several crops. It is also used to control house hold insects and mosquito larvae. It is formulated as 1.5% dust, 20% EC, 50% EC and 10% granules. The dust formulation is marketed under the trade names 'Meesho', 'India Mart', 'Hindol', 'Capban', 'Chloro' etc. 20% EC is marketed under the trade names 'Krishan', 'Starban', 'Dursban', 'Megaban', 'Deviban', 'Tricel 20', 'Chloroban' etc. 50% EC is marketed under the trade names 'Terminator', 'Chlorocon', 'Chloroxa-50', 'Counter-50', 'Chloro-50' etc. The granular formulation is marketed under the trade names 'Deviban 19G', 'Suldrin', 'Hilban' etc. The dust formulation is applied at the rate of 10 kg. per acre. 20% EC is applied at the rate of 400 ml. as spray. 50% EC is applied at the rate of 300 ml. per acre as spray. The granules are applied at the rate of 10 kg. per acre.

ii) **Ethion ($C_9H_{22}O_4P_2S_4$).** Technically it is 'S, S - methylene bis (O, O - dimethyl phosphorodithioate)'. It is a broad spectrum, organo phosphorus, non - systemic, contact and stomach poison effective against both sucking and chewing type of insect pests. It is an insecticide, acaricide and termicide, having quick knock down effect on the insects and it has ovicidal, larvicidal and adulticidal properties. Its residual toxicity persists for a long time on the plant parts. It is effective against Lepidopteran larvae, boll worms, pod borers, shoot borers, white flies, thrips, beetles, scale insects and also mites infesting various crops such as, cotton, rice, chilli, soybean, gram, pigeon pea, tea etc. It is safe for beneficial insects. It inhibits the production of the enzyme acetyl cholinesterase resulting in the blockage of transmission of nerve impulses, thus causing paralysis and ultimate death of the insects. It is formulated as 50% EC and marketed under the trade names 'Ethohit', 'Fosmite', 'Esoc', 'Tefethion', 'Kingmite', 'Hilmite' etc. It is applied at the rate of 400 ml. per acre as spray.

iii) **Malathion ($C_{10}H_{19}O_6PS_2$).** Technically it is 'diethyl (dimethoxy thio-phosphorylthio) succinate - S - 1, 2 - bis (ethoxycarbonyl) ethyl - 0, 0 - dimethyl phosphorodithioate'. It is a non - systemic, broad spectrum, organophosphate insecticide and acaricide with low mammalian toxicity. It is a contact and stomach poison. It acts as an acetyl cholinesterase inhibitor resulting in the blockage of transmission of nerve impulses, thus causing paralysis and ultimate death of the insects. It is effective in controlling a wide range of insect pests such as, rice hispa, pod borer, leaf weevils, *Pyrilla*, aphids, jassids, thrips, white flies and a few caterpillar pests infesting various crops such as, rice, sugarcane, cotton, oilseeds, vegetables, fruit trees, soybean, peas etc. It is also used in the Malaria eradication program for the destruction of mosquitoes. It is formulated as 5% DP and marketed under the trade names 'Growhit', 'K-thion', 'Capalaon', 'Shri Ram Malathion' 'Devimalt' etc. and as 50% EC and marketed under the trade names 'Cythion', 'Mal-50', 'Suthion', 'Chamethion', 'Malathion-60', 'Sandoz 50', 'Hilthion' etc. 5% DP is applied at the rate of 10 kg. per acre as dust and 50% EC is applied at the rate of 250 ml. per acre as spray.

iv) **Methyl Parathion ($C_8H_{10}NO_5PS$).** Technically it is 'O, O - dimethyl - O - 4 - nitrophenyl phosphorothioate'. It is a non - systemic, contact and stomach insecticide and acaricide with some respiratory action. It is a broad spectrum insecticide and is capable of controlling a wide range of insect pests such as, stem borer, leaf folder, hispa, thrips, aphids, green leaf hopper, cut worms and caterpillars infesting several crops such as, rice, wheat, cotton, mustard, coffee etc. It acts by inhibiting the acetyl cholinesterase thereby interferes with the normal transmission of nerve fibres, leading to paralysis and ultimate death of the insects. It is less toxic to mammals. It is formulated as 10% DP and marketed under the trade names 'Missile', 'Paradol', 'Devithion', Kildot' etc., and as 50% EC and marketed under the trade names 'Folidon', 'Devithion', 'Kildot 50E' etc. 10% DP is applied at the rate of 10 kg. per acre as dust and 50% EC is applied at the rate of 300 ml. per acre as spray

v) **Phenthoate ($C_{12}H_{17}O_4PS_2$).** Technically it is 'S - alpha - ethoxycarbonyl benzyl - 0, 0 - dimethyl phosphorodithioate'. It is a broad spectrum, non systemic, organophosphorus insecticide. It is a contact and stomach poison and is capable of controlling a wide range of sucking and chewing insect pests such as, boll worms, case worm, Bihar hairy caterpillar, pod borer, leaf folder, thrips, white flies etc. infesting a large number of plants such as, rice, cotton, groundnut, cardamom, vegetables, green gram, black gram, pulses etc. It has got some ovicidal action also, because of its pungent smell, which deters the moths from laying eggs. However, it is phytotoxic on some varieties of grapes and apples and moderately toxic to mammals. It is a nerve poison and acts as a acetyl cholinesterase inhibitor. It is formulated as 50% EC and marketed under the trade names 'Phendal', 'Royal 50', 'Cidial 50', 'Pestanal', 'Salvo 50' etc. It is applied at the rate of 400 ml. per acre as spray.

vi) **Quinalphos ($C_{12}H_{15}N_2O_3PS$).** It is technically 'O, O - diethyl - O - quinoxabinyl - 2 - yl - phosphorothioate'. It is a broad spectrum, non - systemic, synthetic organophosphorus insecticide and acaricide having contact and stomach action and is capable of controlling a wide range of insect pests such as, stem borer, leaf folder, leaf miner, green leaf hopper, hispa, boll worms, shoot and fruit borer, lady bird beetles etc. infesting several crops such as, rice, cotton, chilli, tomato, groundnut, mustard, soybean, okra, egg plant, black gram, Bengal gram, red gram, sesame etc. It is a nerve poison and acts as an acetyl cholinesterase enzyme inhibitor and causes paralysis and ultimate death of the insects. It is compatible with most of the commonly used plant protection chemicals. It is highly toxic to fishes and other aquatic organisms. It is formulated as 1.5% DP and marketed under the trade names 'Molquin', 'Phoschem', 'Quinalphos 1.5 DP' etc. and as 25% EC and marketed under the trade names 'Krush', 'Ekalux', 'Goldlux', 'Dhanulux', 'Rambalux', 'Hitalux' etc. 1.5% DP is applied at the rate of 10 kg. per acre as dust and 25% EC is applied at the rate of 500 ml. per acre as spray.

vii) **Triazophos ($C_{12}H_{16}N_3O_3PS$).** It is technically 'O, O - diethyl - O - Phenyl - 1 H - 1, 2, 4 - triazol - 3 - yl) - thiophosphate'. It is a broad spectrum, non- systemic, synthetic organophosphorus insecticide and acaricide with some nematicidal action. It is a contact and stomach insecticide and has translaminar action and can penetrate deep into the plant tissues. It is capable of controlling a wide range of sucking and chewing insect pests such as, pink boll worm, spotted boll worm, stem borer, leaf folder, leaf miner, girdle beetle, cut worm, caterpillars, thrips, brown plant hopper, green leaf hopper, white - backed hopper, white flies etc. infesting several crops such as rise, cotton, groundnut, vegetables, soybean, chilli etc. It inhibits the activity of the enzyme acetyl cholinesterase essential for nerve impulse transmission leading to hyper excitement, paralysis and death. It is compatible with most of the commonly used plant protection chemicals. It is formulated as 20% EC and marketed under the trade names 'Shooter', 'Jane' etc. and as 40% EC and marketed under the trade names 'Shook', 'Fulstop', 'Rider', 'Growthion' etc. 20% EC is applied at the rate of 500 ml. per acre as spray and 40% EC is applied at the rate of 250 ml. per acre as spray.

## Organo phosphorus compounds (Systemic)

i) **Acephate ($C_4H_{10}NO_3PS$).** Technically it is - 'O, S - dimethyl acetyl phosphramido thioate'. It is a systemic insecticide with moderatepersistence for 10 - 15 days. It is effective in controlling sucking insect pests such as, aphids, thrips, hoppers, white flies, mealy bugs etc. as well as biting and chewing pests such as, Lepidopterous larvae, leaf folders, stem borers, boll worms, army worms, fruit borers etc. attacking several crops such as, rice, cotton, potato, tomato, sugarcane, vegetables, tobacco, fruit crops etc. In the insecticide applied fields, it is converted by insects into **'methamidophos'**, which is a more potent and highly mobile insecticide. It affects the central nervous system of insects and causes their death. It is formulated as 75% WSP and marketed under the trade names 'Acephate' 'Acetox', 'Acataf', 'Orthene', 'Starthene', 'Lancer', 'Ace', etc. and is applied at the rate of 300 - 400 g. per acre as spray. It is also formulated as 95% SG and marketed under the trade names 'Hunk', 'Starphate Super' etc. It is applied at the rate of 150 - 300 g. per acre as spray.

ii) **Chromafenozide ($C_{24}H_{30}N_2O_3$).** It is a systemic, contact and stomach poison effective against larvae of Lepidopteran insect pests. It is effective at all stages of larval development and causes cessation of feeding and movement of insects. It inhibits production of the hormone ecdysone, thereby affects the moulting process of larvae. It is very effective against leaf folders and stem borers of crops such as, rice, sorghum, sugarcane etc. It is safe to natural enemies of insect pests. It is formulated as 80% WP and marketed under the trade names 'Dodger', 'India Mart', 'Matrix' etc. It is applied at the rate of 50 g. per acre as spray

iii) **Dimethoate ($C_5H_{12}NO_3PS_2$).** Its technical name is 'O, O - dimethyl S - (N - methyl cabomoylmethyl) Phosphorothiothioate'. It is a broad spectrum, systemic, contact insecticide and acaricide. It is highly effective in controlling a wide range of sucking insect pests such as, white flies, aphids, thrips, beetles, weevils and mites of a wide variety of crops, as well as lice attacking poultry. It is an acetyl cholinesterase inhibitor, which inhibits production of cholinesterase, an enzyme vital for the functioning of the nervous system and causes death of the insect. One of its breakdown product is **'Omethoate'**, a potent cholinesterase inhibitor, ten times more toxic than its parent compound. Its residual toxicity persists for many days on the plant parts. It is phytotoxic to some varieties of sorghum and chrysanthemum. It is highly compatible with most of the other insecticides and fungicides. It is formulated as 30% EC and marketed under the trade names 'Rogor', 'Cygon', 'Hexagor', 'Sagar' etc. It is applied at the rate of 200 ml. per acre as spray.

**iv) Monocrotophos ($C_7H_{14}NO_5P$).** It is technically 'Dimethyl (E) - I - methyl - 2 - (methylcarbamoyl) vinyl phosphate'. It is a broad spectrum, organophosphorus, systemic insecticide and acaricide. It is a contact and stomach poison with long residual action. It is effective in controlling a wide range of sucking and chewing insect pests such as, Lepidoperous larvae, boll worms, aphids, thrips, mealy bugs, white flies, green leaf hopper, brown plant hopper, weevils, shoot fly, pod borer, coconut black - headed caterpillar, mites etc. infesting a large number of plants such as, rice, maize, sugarcane, cotton, coffee, cardamom, mango, coconut, black gram, green gram, red gram etc. Monocrotophos affects the nervous system of insects by inhibiting acetyl cholinesterase an enzyme essential for the normal transmission of nerve impulses, which causes paralysis and ultimate death of the insect. It is highly toxic to man and mammals and its residual toxicity persists in the plant parts for a long time. Hence its use has been banned from use on vegetable crops and greens. It is formulated as 36% EC and marketed under the trade names 'Phoskil', 'Monokil', 'Monorin', 'Monster', 'Monoxil', 'Luphos', 'Monophos', 'Monogreen', 'Maha Vinash' etc. It is applied at the rate of 250 ml. per acre as spray.

**v) Profenophos ($C_9H_{22}O_4P_2S_4$).** Technically it is '0 - (4 - Bromo - 2 - chlorophenyl) - 0 - ethyl - S - propyl phosphorothioate'. It is a broad spectrum, organophosphorus, systemic insecticide and acaricide. It is a contact and stomach poison with long lasting residual activity. It is effective in controlling a wide range of sucking and foliar feeding larvae such as, caterpillars, aphids, thrips, leaf hoppers, white flies, all kinds of mites etc. infesting crops such as, rice, maize, cotton, soybean, chilli, potato, tobacco, tea, vegetables, fruit trees etc. It should not be used on cucurbits, as it causes phytotoxicity. It is a neurotoxin and acts as an acetyl cholinesterase enzyme inhibitor, which disrupts normal transmission of nerve impulses resulting in paralysis and ultimate death of the insects. It is toxic to birds, honeybees, fishes and other aquatic organisms, as well as to small mammals. It is compatible with most of the commonly used plant protection chemicals. It is formulated as 50% EC and marketed under the trade names 'Prahar', 'Pro Gold', 'Curacron', 'Profen', 'Jashn', 'Promax', 'Pro Kill', 'Pestanal' etc. It is applied at the rate of 300 ml. per acre as spray.

## Organofluorine compounds

i) **Flubendiamide ($C_{23}H_{22}F_7IN_2O_4S$).** It is a non - systemic, broad spectrum, contact and stomach insecticide, effective in controlling a wide range of chewing and sucking insect pests, especially the Lepidopterous larvae. It controls different types of caterpillars, stem borers, leaf roller, American cotton boll worm, spotted boll worms of cotton, pod borers, fruit borers, diamond back moth etc. infesting various crops such as, rice, cotton, chilli, tomato, egg plant, cabbage, black gram, bengal gram, soybean, pigeon pea etc. It affects the nervous system of insects, immobilizes the insect muscles inducing cessation of feeding and causing ultimate death. It is formulated as 20% WG and marketed under the trade names 'Takibi', 'Tata Takumi' etc. and as 39.37% SC and marketed under the trade names 'Fame', 'Fluben', 'Sperzite', 'Flubensik' etc. The water soluble granule is applied at the rate of 100 g. per acre as spray and the soluble concentrate is applied at the rate of 50 ml. per acre as spray.

**Carbamates**. The insecticidal carbamates are derivatives of carbamic acid and dithiocarbamic acid. Insecticidal carbamates, like organophosphorus insecticides, are also cholinesterase inhibitors. Carbamate poisoned insects show violent convulsions and other neuromuscular disturbances leading to their ultimate death.

## Carbamates (Non - systemic)

i) **Fenobucarb ($C_{12}H_{17}NO_2$).** Its technical name is '2 - (butan - 2 - yl phenyl methyl carbamate'. It is a broad spectrum, stomach and mainly contact, non - systemic, synthetic carbamate insecticide. It controls a wide range of sucking insect pests and a few other biting insect pests such as, green leaf hopper, brown plant hopper, mango hopper, jassids, thrips, white flies, leaf rollers, stem borers etc. infesting various crops such as, rice, cotton, sugarcane, vegetables etc. It is toxic to fishes and water inhabiting organisms, but harmless to beneficial insects. It inhibits the production of the enzyme acetyl cholinesterase and affects the nervous system resulting in paralysis and ultimate death of the insects. It is formulated as 50 % EC and marketed under the trade names 'Wardon', 'Pestanol', 'Lawin', 'Pinaka', 'Win Win' etc. It is applied at the rate of 300 ml. per acre as spray.

ii) **Thiodicarb ($C_{10}H_{18}N_4O_4S_3$).** Technically it is '1 - (methylsulfanyl ethylidone) amino - N - methyl - N - (3 - methyl - 6 - oxa - 2 - thia - 4 - 7 - diazaoct - 3 - en - 7 - yl) sulfanyl) carbamate'. It is a broad spectrum, non - systemic carbamate insecticide and molluscicide. It is a strong stomach poison and has some contact action. Primarily it acts as an ingestion toxicant and also acts as an acetyl cholinesterase enzyme inhibitor. It exhibits ovicidal, larvicidal and adulticidal action and has long residual toxicity. It is effective in controlling a wide range of insect pests such as, boll worms, diamond - back moth, weevils, beetles, shoot borer, pod borer, loopers, fruit borer, especially gregarious polyphagous leaf feeders, such as *Spodoptera,* and *Helicoverpa*, slugs and snails etc. infesting various crops such as, cotton, chilli, cabbage, egg plant, black gram, green gram, tomato, tobacco etc. It is formulated as 75% WP and marketed under the trade name 'Larvin'. It is applied at the rate of 300 gm. per acre as spray.

**Carbamates (Systemic)**

i) **Carbofuran ($C_{12}H_{15}NO_3$).** Its technical name is (2, 3 - dihydro - 2, 2 - dimethyl - 7 - benzofuranyl). It is a broad spectrum contact and stomach poison and a systemic insecticide and nematicide . It controls a wide range of foliar pests, soil pests and a few nematodes. Stem borers of maize, sugarcane, rice etc. are controlled by this insecticide. It is also effective in controlling sucking pests, thrips, mites and soil inhabiting pests such as, rice root weevils, maize root worms, white grubs, flea beetle larvae and fly maggots. It blocks nerve transmission of impulses by inhibiting acetyl cholnesterase resulting in paralysis and ultimate death of the insects. Carbofuran application stimulates crop growth in cotton, rice, tobacco, sorghum and maize. It is mainly formulated as 3% granules and marketed under the trade names 'Furadan 3G', 'Carbogran 3G', 'Vinfuran 3G', 'Furacarb 3G', 'Curater 3G' etc. It is applied either to the soil at 13 kg. per acre or as whorl application.

ii) **Carbosulfan ($C_{20}H_{22}N_2O_3S$).** Its technical name is '2, 3 - dihydro - 2, 2 - dimethylbenzofuran - 7 - yl (dibutylaminothio) - methyl carbamate'. It is a broad spectrum, systemic insecticide effective against a wide range of sucking and chewing pests of many crops. In the sprayed plants, it is metabolized into **'carbofuran'** and **'3 - hydrocarbofuran'**, which are also insecticidal. It is formulated as 25% EC and is marketed under the trade names 'Marshal', Aatank', 'Carbosulfan' etc. It is applied at the rate of 400 ml. per acre as spray.

iii) **Methomil ($C_5H_{10}N_2O_2S$).** Technically it is 'S - methyl - N - (methylcarbomoyloxy) thiacetimidate'. It is a broad spectrum, systemic, synthetic carbamate, contact and stomach insecticide. It has quick, knock - down effect. It is ovicidal and larvicidal. It acts as an acetyl cholinesterase inhibitor that affects the nervous system of insects and cause their ultimate death. It is capable of controlling a wide range of insect pests such as, leaf feeding caterpillars, cut worm, army worm, cotton boll worms, fruit borers, pod borers, leaf miners, thrips, mealy bugs etc. infesting several crops such as, rice, cotton, tomato, pigeon pea, chilli, sorghum, vegetables, egg plant etc. It is compatible with most of the common plant protection chemicals. It is formulated as 40% SP and marketed under the trade names 'Dragon', 'Dash', 'Atom-40', 'Lannate', 'Kinet' etc. It is applied at the rate of 300 g. per acre as spray.

**Neonicotinoids (Systemic)**

i) **Acetamiprid ($C_{10}H_{11}Cl\ N_4$).** Technically it is 'N - (6 - Chloro - 3 -Pyridinyl) methyl- N - cyano - N - Methyl - ethanimidamide'. It is a systemic insecticide similar to nicotine and has high residual toxicity. It is highly effective in controlling sucking pests such as, aphids, jassids, thrips, hoppers etc., as well as fruit moths, leaf miners, plant bugs etc. It is effective in controlling pests, which have developed resistance to other insecticides. It is a nervous poison and has ovicidal, larvicidal and adulticidal activity. It is also safe to natural enemies of insect pests. It is mainly formulated as 20% SP and marketed under the trade names 'Manik', 'Ikon', 'Baadshah', 'Prime Gold', 'Proud', 'Dhanpreet', 'Acepro', 'Active', 'Acelon' etc. It is applied at the rae of 40 - 80 g. per acre as spray.

ii) **Clothianidin ($C_6H_8ClN_5O_2S$).** It is technically '1 - (2 - chloro - 1, 3 - thiozol - 5 - yl methyl) - 3 - methyl - 2 - nitroguanidine'. It is a systemic, neonicotinoid, broad spectrum, contact and stomach poison. It is very effective in controlling sucking pests such as, brown plant hopper, jasssids, mites, white flies, mealy bugs, mosquito bugs etc.of many crops such as, rice, cotton, sugarcane, tea, as well as chewing pests. It affects the central nervous system of insects and causes their death. It is formulated as 50% water dispersible granules and marketed under the trade name 'Dantotsu'. It is applied at the rate of 60 g. per acre as spray

iii) **Dinotefuran ($C_7H_{14}N_4O_3$).** Technically it is 'N - methyl - N - nitro - N - (tetrahydro - 3 - furanyl) - methyl - guanidine'. It is a highly systemic, contact and stomach insecticide belonging to the neonicotinoid class insecticide. It affects the central nervous system leading to over stimulation and consequent paralysis and ultimate death of the insects. It has got quick knock down effect and has long residual toxicity. It controls a wide range of insect pests such as, brown plant hopper, jassids, aphids, thrips, white flies, mealy bugs, diamond back moth etc. infesting various crops such as, rice, cotton, cucurbits, onion, tomato, vegetables etc. It is formulated as 20% SG and marketed under the trade names 'Dinotrex', 'Osheen', 'Oshin', 'Nowko', 'Token' etc. It is applied at the rate of 120 g. per acre as spray.

iv) **Imidacloprid ($C_9H_{10}ClN_5O_2$).** Its technical name is '1 - (6 - chloro - 3 - pyridinyl) - methyl - N - nitro - 2 - imidazolidinimine'. It is a broad spectrum, highly systemic, contact and stomach insecticide belonging to the neonicotinoids. It is fast acting and has quick knock down effect on the insects. It affects the central nervous system, causing paralysis and ultimate death of the insects. It is quite effective against a wide range of sucking insect pests such as, aphids, jassids, brown plant hopper, green leaf hopper, white - backed hopper, mango hopper, thrips, leaf miners, some beetles etc. infesting various crops such as, rice, cotton, sugarcane, chilli, tomato, egg plant, lady's finger, mango etc., It is compatible with most of the commonly used insecticides and fungicides and it does not affect the environment. It is formulated as 17.8% SL and marketed under the trade names 'Media', 'Green mida', 'confidor', 'Imidacel', 'Imida Gold' etc., as 30.5% SC. and marketed under the trade names 'Green Mida', 'Isogashi', 'Tatamida', 'Hi-imida', 'Media', 'Confidor Super', ':Protect', Jumbo' etc. and as 70% WG and marketed under the trade names 'Admit', 'Admire', 'Bomba', Ad-fyre', 'IMD-70' etc. Imidacloprid 17.8 SL is applied at the rate of 100 ml. per acre as spray, 30.5 SC is applied at the rate of 50 ml. per acre as spray and 70% WG at 15 gm. per acre as spray. Imidachloprid is also formulated as a seed treating insecticide and marketed under the trade names 'Remix', 'Guaoho', 'Aakrosh', ' Bajiao', 'Farmida', 'Emidox' etc.. The seedlings from the treated seeds of crops such as, cotton, sunflower, sorghum, pearl millet, soybean, lady's finger etc. remain free from sucking pest infestation for a period of 30 - 40 days after germination. For seed treatment Imidachloprid 48% SL/FS is applied at the rate of 0.5 - 0.9 ml. per kg. of seeds.

v) **Thiacloprid ($C_{10}H_9ClN_4S$).** Technically it is '3 - (6 - chloro - 3 - pyridimylmethyl) - 2 - thiazolidimylidene - cynamide'. It is a broad spectrum, systemic, contact and stomach insecticide belonging to the neonicotinoid group of insecticides. It is effective in controlling a wide range of insect pests such as Lepidopterans, aphids, thrips, jassids, white flies, stem borer, leaf folder, shoot and fruit borer, mosquito bug, girdle beetle etc. infesting various crops such as, rice, cotton, chilli, brinjal, soybean, tea, apple etc. It is quick acting and has knock - down effect on sucking and chewing insect pests. It is effective on insect pests, which have developed resistance to other insecticides. It is antagonistic to nicotinic acetyl choline receptor in the central nervous system and disturbs proper signal transmission leading to excitement, paralysis and ultimate death of the insects. It is formulated as 21.7% SC and marketed under the trade names 'Alanto', 'Gunwaar', 'Splendour', 'Scil' etc. It is applied at the rate of 200 ml. per acre as spray.

vi) **Thiamethoxam ($C_8H_{10}ClN_5O_3S$).** Technically it is '3 - (2 chloro - 5 - thiazolyl methy) - tetrahydro - 5 - methyl - N - nitro - 4H - 1, 3, 5 - oxadiazine - 4 - imine'. It is a broad spectrum, highly systemic, fast acting and long lasting, contact and stomach insecticide with some repellent action. It belongs to the neonicotinic group of insecticides. It is effective in controlling a wide range of sucking and chewing insect pests such as, stem borer, leaf folder, gall midge, brown plant hopper, white - backed plant hopper, green leaf hopper, thrips, aphids, white flies, plant hoppers, mealy bugs, mosquito bugs, ash weevils, psyllids etc. infesting several crops such as, rice, cotton, chilli, wheat, groundnut, mustard, tomato, tobacco, vegetables, potato, brinjal, tea, coffee, apple etc. It acts on the nicotinic acetyl cholin receptor of the nervous system and inhibits the feeding reflex of insects, as such the insects die of starvation. It is compatible with most of the commonly used pesticides. It is formulated as 25% WG and marketed under the trade names 'Actara', 'Pestanal', 'Arava', 'Thioxam', 'Anant', 'Theme', 'Click', 'Evident', 'Eloxa' etc. and as 75% SG and marketed under the trade names 'Capsadis', 'Shutter', 'Devasena', 'Thioxam Super' etc. 25% WG is applied at the rate of 80 gm. per acre as spray and 75% SG is applied at the rate of 60 gm. per acre as spray. Thiamethoxam 30% FS marketed under the trade names 'Areva Super', 'Tracker', 'Kaziro', 'Tata Trot', 'Reno' etc. and Thiamethoxam 70% WS marketed under the trade names 'Cruiser', 'Texan', 'Thia Gold', 'Taxan' etc. are used as seed dressing insecticides to control the pests, which attack the germinating seedlings of crops such as, cotton, sunflower, rice, pulses etc.

## Synthetic pyrethroids

i) **Alpha Cypermethin ($C_{22}H_{19}C_{12}NO_3$).** It is technically - 'Cyano (3 - phenoxyphenyl) Methyl 3 - (2, 2 dichloroethenyl) - 2-2 - Dimethyl cyclopropane carboxylate'. It is a non - systemic, synthetic pyrethroid insecticide. It is a contact and stomach poison and controls a wide range of chewing and sucking insect pests of several crops such as, fruits, vegetables, vines, cereals, oilseeds, cotton, rice etc. It mainly acts on the nervous system and gives a quick knock down action. However it is toxic to fishes and bees. It is mainly formulated as 10% EC and marketed under the trade names 'Alfastar', 'Guru', 'Karfu', 'Safari', 'Dash' etc. It is applied at the rate of 100 - 125 ml. per acre as spray.

ii) **Alphamethrin ($C_{22}H_{19}Cl_2NO_3$).** It is a broad spectrum, non - systemic, synthetic pyrethroid contact and stomach poison, which acts mainly on the nervous system of target pests and has quick knock down action. It is effective against both chewing and sucking insect pests of several crops. It is mainly formulated as 10% EC and marketed under the trade names 'Mig 10', 'Grand 301', 'Tata Alpha', 'Gem', 'Thrill', 'Alfa', 'Alfa Gold' etc. It is applied at the rate of 100 ml. per acre as spray.

iii) **Bifenthrin ($C_{23}H_{22}ClF_3O_2$).** It is a broad spectrum, non - systemic, synthetic pyrethroid contact and stomach poison, which acts mainly on the nervous system of the target pests and has quick knock down action. It is effective in controlling both chewing and sucking type of insect pests attacking several crops. It is highly toxic to fishes and bees. It is mainly formulated as 10% EC and marketed under the trade names 'Hectastar', 'Rock', 'Metastar', 'Markar', 'Super Star', 'Canister', 'Centrix' etc. It is applied at the rate of 200 ml. per acre as spray.

iv) **Cypermethrin ($C_{22}H_{19}Cl_2NO_3$).** Technically it is 'Cyano (3-phenoxyphenyl) methyl - 3 - (2, 2 dichloroehenyl) - 2, 2 - dimethylcyclopropane - 1 - carboxylate'. It is a non - systemic, broad spectrum, contact and stomach poison effective against a wide range of sucking and chewing insect pests infesting many crops. It affects the nervous system of insects by blocking the nerve impulses resulting in their death. It is formulated as 10% EC and marketed under the trade names 'Ripcord', 'Challenge', 'Lacer', 'Cypermar', 'Bilcyp', 'Polytrin' etc. and as 25% EC. and marketed under the trade names 'Cymbush', 'Sandoz Cyperkil', 'Cypervip' etc. Cypermethrin 10% is applied at the rate of 100 ml. per acre as spray and Cypermethrin 25% is applied at the rate of 60 ml. per acre as spray.

v) **Fenpropathrin ($C_{22}H_{23}NO_3$).** Technically it is 'Cyano - (3 - Phenoxyphenyl) methyl - 2, 2, 3, 3 - tetramethylcyclopropane - 1 - carboxylate'. It is a non - systemic, synthetic, broad spectrum pyrethroid insecticide and acaricide. It is effective in controlling a large number of pests such as, aphids, jassids, hoppers, Army worm, boll worms, bud worm, white flies, loopers, diamond back moth, fruit moths, leaf roller etc., as well as mites. infesting various crops such as, cotton, chilli, egg plant, lady's finger, tea, rice etc. It alters the nerve functioning and causes paralysis in insects and ultimate death. It is formulated as 30% EC and marketed under the trade name 'Meothrin'. It is applied at the rate of 100 ml. per acre as spray.

vi) **Fenvalerate ($C_{25}H_{22}ClNO_3$).** Its technical name is 'Cyano (3 - phenoxy phenyl) methyl - 4 - chloro - alpha - (1 - methyl ethyl) benzene acetate'. It is a non - systemic, synthetic, broad spectrum, pyrethroid, contact and stomach poison. It controls a wide range of chewing and sucking insect pests such as, aphids, jassids, Army worm, boll worms, bud worms, fruit borers, diamond back moth etc. infesting various crops such as, groundnut, cotton, tomato, red gram, etc. It affects the peripheral and central nervous system of insects, causes paralysis leading to quick death. It is formulated as 0.4% DP and marketed under the trade names 'Fielder', 'Anufen', 'Fenoxx', 'Ghatak' etc. and as 20% EC. and marketed under the trade names 'Tatafen', 'Sumicidin', 'Fenhit', 'Fenvalerate' etc. The dusting powder is applied at the rate of 10 kg. per acre and the emulsifying concentrate is applied at the rate of 150 - 200 ml. per acre as spray.

vii) **Deltamethrin ($C_{22}H_{19}Br_2NO_3$).** Technically it is '(S) - alpha - cyano - 3 - phenoxybenzyl (1R, 3R) - 3 - (2 - 2 - dibromovinyl) - 2 - 2 dimethyl cyclopropane carboxylate'. It is a non - systemic, synthetic pyrethroid, broad spectrum insecticide. It is a contact and stomach poison, effective against both sucking and chewing type of insect pests such as, aphids, thrips, boll worms, fruit borers, leaf folders, stem borers etc. infesting various crops such as, rice, cotton, tomato, lady's finger, tea, tomato, chilli, onion etc. It affects the nervous system of insects and disrupts the conduction of nervous impulses, causing paralysis and ultimate death of the insects. It has quick, knock-down effect on the insects. It is formulated as 2.8% EC and marketed under the trade names 'Decis', 'Del 25', 'Redox', 'Delya Hit', 'Ecdel 28' etc. and as 11% EC and marketed under the trade names 'Derin', 'Delta 11', 'Tokiton', 'Delta Super' etc. 2.8% EC is applied at the rate of 200 ml. per acre as spray and 11% EC is applied at the rate of 60 ml. per acre as spray.

viii) **Etofenprox ($C_{25}H_{28}O_3$).** Technically it is '2 - (4 - ethoxyphenyl) - 2 - methylpropyl - 3 - phenoxybenzyl ether'. It is a non - systemic, broad spectrum, synthetic pyrethroid insecticide. It is a contact and stomach poison, effective against both sucking and chewing insect pests. It controls a wide range of insect pests such as, green leaf hopper, brown plant hopper, stem borers, leaf folders, whorl maggot, aphids, caterpillars, fruit borers etc. infesting several crops such as, rice, cotton, fruits, apple, vegetables, soybean, tea, cereals etc. It has got quick, knockdown effect and has long residual toxicity. Just as other synthetic pyrethroids, this insecticide also affects the nervous system by disrupting the electrical signaling impulses, causing paralysis and ultimate death of the insects. It is toxic to fishes and honey bees. It is formulated as 30% EC and marketed under the trade names ' Dash', 'Trebon', 'Etofenprox' etc. It is applied at the rate of 100 ml. per acre as spray.

ix) **Fenpropathrin ($C_{22}H_{23}NO_3$).** Technically it is 'cyano(3-phenoxyphenyl) methyl - 2, 2, 3, 3 - tetramethyl cyclo propane carboxylate'. It is a broad spectrum, non - systemic, synthetic pyrethroid insecticide and acaricide. It is a contact and stomach poison and capable of controlling both sucking and biting insect pests. It is effective in controlling a wide range of insect pests such as, aphids, jassids, hoppers, white flies, Army worm, boll worms, bud worms, loopers, stem borers, tuber worms, mosquitoes etc. infesting several crops such as, rice, cotton, cabbage, egg plant, lady's finger, tea etc. It interferes with the functioning of the sodium channels of the nervous system resulting in paralysis and ultimate death of the insects. It is formulated as 10% EC and marketed under the trade name 'Danitol' and as 30% EC and marketed under the trade name 'Meothrin'.10% EC is applied at the rate of 300 ml. per acre as spray and 30% EC is applied at the rate of 100 ml. per acre as spray.

x) **Hexthiazox ($C_{17}H_{21}ClN_2O_2S$).** Technically it is 'Trans-5-(chlorophenyl) - N - cyclohexyl - 4 - methyl - 2 - ozo - 3 - thiazolidin carboximide'. It is a broad spectrum, non - systemic, contact and stomach acaricide. It has got ovicidal, larvicidal and nymphicidal properties. It acts as a insect growth regulator and inhibits moulting in the early stages of development and thus prevents growth and causes death of the insects. It is effective against a wide range of sucking insect pests, specially mites infesting various crops such as, cotton, chilli, tea, apple, grapes, rose, egg plant, okra etc. It is safe to the environment and harmless to beneficial insects and natural enemies of insect pests. It is compatible with all the commonly used insecticides and fungicides. It is formulated as 5.45 EC and marketed under the trade names 'Maiden', 'Cubax', 'Miteban', 'Mitolin' etc. It is applied at the rate of 200 ml. per acre as spray.

xi) **Lambda cyhalothrin ($C_{23}H_{19}ClF_3NO_3$).** Technically it is 'Cyano (3 - phenoxyphenyl) methyl - 3 (2 - chloro - 3, 3, 3 - trifluoro - 1 - prophenyl) 2 - dimethylcyclopropane carboxylate'. It is a broad spectrum, non systemic, synthetic pyrethroid pesticide. It is a contact and stomach poison and has repellent action also. It has quick, knock - down effect and its residual toxicity lasts for a long time. It is effective in controlling a wide range of insect pests such as, cotton pink boll worm, spotted boll worms, jassids, thrips, leaf folder, green leaf hopper, stem borers, leaf folder, gall midge etc. infesting various crops such as cotton, rice, cereals, potato, tomato, tobacco, vegetables etc. It acts as a neurotoxin that targets the sodium channels in the membranes of neurons in the central nervous system, causes paralysis and ultimate death. It is highly toxic to honeybees, fishes and other aquatic organisms. It is compatible with most of the plant protection chemicals. It is formulated as 2.5% EC and marketed under the trade names 'Kozuka', 'Reeva', 'Lamdex', 'Karate', 'Loxa', 'Golden Dragon' etc., as 4.9% CS and marketed under the trade names 'Metador', 'Yuri', 'Attack', 'Lakra', 'Jackpot', 'Elthrin Super', 'Killer', 'Mantra' etc., and as 5.0% EC and marketed under the trade names 'Lambrada', 'Lamdex Gold', 'Karate', 'NCON', 'Lambda Super Green', 'Clomda - 5', 'Singham' etc. 2.5% EC is applied at the rate of 500 ml. per acre as spray, 4.9%CS is applied at the rate of 300 ml. per acre as spray and 5.0% EC is applied at the rate of 300 ml. per acre as spray.

xii) **Permethrin ($C_{21}H_{20}Cl_2O_3$).** Technically it is '3 - phenoxybenzyl (1RS, 3RS, 1RS, 3SR) - 3 - (2, 2 dichloroethenyl) - 2 - 2 - dimethyl cyclopropane carboxylate'. It is a broad spectrum, non - systemic, synthetic pyrethroid insecticide. It is a contact and stomach poison with mild repellent action. It is effective in controlling a wide range of insect pests such as, boll worms, leaves, flowers and fruit eating pests, thrips, pod borer, early shoot borer, diamond back moth, mites, aphids, jassids, brown plant hopper, green leaf hopper, white flies etc. infesting various crops such as, rice, cotton, chilli, sugarcane, cabbage, grapes, red gram, lady's finger, chick pea etc. It acts on the nervous system of insects. It interferes with the sodium channels and disrupt the functions of neurons and causes muscle spasms culminating in paralysis and death. It is highly toxic to honeybees and aquatic organisms. It is compatible with most of the commonly used plant protection chemicals. It is formulated as 25% EC and marketed under the trade names 'Pixel 25', 'Signor', 'Proud', 'Brahmastra', 'Agniban', 'Hawk', 'Tag bush' etc. It is applied at the rate of 400 ml. per acre as spray.

## Thiazine group

i) **Buprofezin ($C_{16}H_{23}N_3O_5$).** It is an insecticide of 'Growth Regulator Group'. It inhibits moulting of nymphs and larvae by hindering the formation of exoskeleton in the nymphs leading to their death. It controls all the nymphal stages of insects and inhibits the egg - laying capacity of female insects. It is highly effective in controlling sucking pests such as, white flies, aphids, jassids, thrips, mites, mealy bugs and hoppers infesting rice, cotton, chillies, grapes, mango etc. It is mainly formulated as 25% EC and marketed under the trade names 'Jawaa', 'Flotis', 'Trust', 'Devifezin', 'Applaud', 'Apple', 'Tribume', 'Ninja' etc. It is applied at the rate of 300 ml. per acre as spray.

## Nereistoxin analogue

i) **Cartap ($C_7H_{16}ClN_3O_2S_2$).** Its technical name is 'S, S - (2 - dimethylamino - trimethylene) bis thiocarbamate'. It is a systemic, contact and stomach poison belonging to the Nereistoxin analogue group of insecticides. It has ovicidal, larvicidal and adulticidal properties and is highly effective in controlling several insect pests of many crops, especially stem borers and leaf miners. It is safe for the beneficial insects and is compatible with all the commonly used insecticides and fungicides. It is formulated as 4.0% granules and marketed under the trade names 'Katsu', 'Dartriz', 'Caldan', 'Marktap', 'Ind Gold', 'Fastap' etc. It is applied to the soil at the rate of 10 kg. per acre. It is also formulated as 50% SP and marketed under the trade names 'Caldan', 'Dartriz', 'Cartap' etc. It is applied as spray at the rate of 400 g. per acre.

## Pyrrole group

i) **Chlorofenapyr ($C_{15}H_{11}BrClF_3N_2O$).** Technically it is '4 - Bromo - 2 - (4 - chlorophenyl) - 1 - (ethoxymethyl) - 5 - (trifluoromethyl) - 1 H - pyrrole - 3 - carbonitrile'. It is a contact and stomach poison and controls a wide range of insect pests including mites, thrips, aphids and several other crop pests. It interrupts the production of 'adenosine triphosphate' in the insect nervous system resulting in their death. It is safe to the beneficial insects and bees. It is formulated as 10% SC and marketed under the trade names 'Intreprid', 'Cutlass', 'Record', 'Lepido' etc. It is applied at the rate of 250 ml. per acre as spray

## Pyrazole group

i) **Fenpyroximate ($C_{24}H_{27}N_3O_4$).** Technically it is '1 - butyl (E) - alpha - (1, 3 -dimethyl - 5 - phenoxypyrazol - 4 - yl methyleneaminoxy) - p - toluate'. It is mainly a contact insecticide, acaricide and miticide highly effective against red spider mite, yellow mite, pink mite, purple mite and eriophyid mite on crops such as, tea, coconut, chilli etc. It has got moulting and oviposition inhibition action and has quick knock down effect against both nymphs and adults. It is safe to honeybees and beneficial insects but highly toxic to fishes. It is formulated as 5.0% EC and marketed under the trade names 'Seagate', 'Pyromite', 'Akari', 'Sadna' etc. It is applied at the rate of 200 ml. per acre as spray.

ii) **Fipronil ($C_{12}H_4Cl_2FeN_4OS$).** Echnically it is '5 - amino - 1 - 2, 6 - dichloro - 4 - (trifluoromethyl sulfinyl) pyrazole - 3 - carbonitrite'. It belongs to the phenyl pyrazole group of insecticides. It is a broad spectrum, highly systemic, contact and stomach poison and capable of controlling a wide range of chewing and sucking insects infesting crops such as, rice, cotton, chilli, cabbage, sugarcane etc. It has got high residual toxicity and can control insect pests, which have developed resistance to other insecticides. It acts on the nervous system of insects by interfering in the nerve impulse transmission. It is highly toxic to fishes and other water inhabiting organisms. It is formulated as 0.3% G and marketed under the trade names 'Recent', 'Farita', 'Shinzer', 'Katyayani', 'Fipro', 'Janbaaz', 'Heranil' etc. and as 2.92% EC and marketed under the trade names 'Agenda', 'Fipronil', 'Bio M Power', 'Chilli KOR' etc. and as 5.0% SC. and marketed under the trade names 'Aashirwaad', 'Shinzon', 'Bell', 'Recely', 'Getter' etc. and as 80% WG and marketed under the trade names 'Jump', 'Fipro 80', 'Joker', 'Ahead' etc. The granular formulation is applied at the rate of 10 kg. per acre. The emulsifiable concentrate formulation is applied at the rate of 400 ml. per acre as spray. The soluble concentrate formulation is applied at the rate of 200 ml. per acre as spray. The wettable granular formulation is applied at the rate of 25 g. per acre as spray.

iii) **Chlorantraniliprole ($C_{18}H_{14}BrCl_2N_5O_2$).** Technically it is '3 - Bromo - 4 - chloro - 1 - (3 - chloro - 2 - pyridyl) - 2 - methyl - 6 - (methylcarbamoyl) - 1 H - pyrazole - 5 - carbaoxamide'. This synthetic insecticide is formulated based on the extract of *Ryania speciosa,* a wild plant, which has got insecticidal properties. It is a wide spectrum, systemic, contact and stomach insecticide and can control a wide range of insects belonging to Lepidoptera, beetles, flies, termites etc. infesting crops such as, rice, cotton, Sugarcane, tomato, pulses, maize, groundnut etc. It affects the muscles, disrupts normal muscle contraction leading to paralysis and ultimate death of the insects. It is quite safe to parasitoids and predators. It is formulated as 0.4% G and marketed under the trade names 'Kashima', 'Fertena' 'Fantasic, 'Vistara, 'Stambh' etc. and as 18.5 SC and marketed under the trade names 'Shimo', 'Coragen', 'Cover' etc. The granules are applied at the rate of 6.0 kg/ acre and the soluble concentrate is applied at the rate of 60 ml. per acre as spray.

**Thiourea group**

i) **Diafenthiuron ($C_{23}H_{30}N_2O_5$).** Technically it is '1 - tert - butyl - 3 - (2, 6 - di - isopropyl - 4 - phenoxyphenyl) thiourea'. It belongs to the thiourea group of insecticides. It is a systemic, broad spectrum, contact and stomach insecticide and has got some vapor action also. It is very effective in controlling sucking pests such as white flies, aphids, jassids, thrips, diamond back moth as well as mites infesting crops such as, cotton, chilli, egg plant, cardamom, cabbage, cauliflower, brocoli, roses etc. It controls both the nymphs and adults effectively. It causes paralysis of the insects and they remain immobile for a few days and die. It has quick, knock down effect. It is safe for the beneficial insects. It is formulated as 50% WP and marketed under the trade names 'Derby', 'Pegasus', 'Shoku', 'Pager', 'Ashwamed' etc. It is applied at the rate of 250 g. per acre as spray.

**Evermectin group**

i) **Emamectin Benzoate ($C_{56}H_{81}NO_{15}$).** It is a macrocyclic lactone insecticide. It is a non - systemic, contact and stomach poison. It is a broad spectrum insecticide and is very effective in controlling insects such as, boll worms of cotton, fruit and shoot borers, thrips etc. infesting crops such as, cotton, tea, grapes, cabbage, lady's finger, egg plant, chilli, red gram, chick pea etc. The active ingredient Evermectin in the insecticide causes paralysis of larvae due to the activation of the chloride channel at the nerve level resulting in the death of the insect. It is formulated as 5% SG and marketed under the trade names 'Commander', 'Jupiter', 'Egro', 'Ema Gold' etc. It is applied at the rate of 200 g. per acre as spray.

**Oxadiazine group**

i) **Indoxacarb ($C_{22}H_{17}ClF_3N_3O_7$).** It is a non - systemic, broad spectrum, contact and stomach poison. It is very effective in controlling Lepidopterans such as, cotton pink boll worm, spotted boll worms, cut worms, army worms, fruit borers, leaf folders, hoppers, thrips etc. infesting crops such as, cotton, chilli, tobacco, tomato, cabbage, lady's finger, black gram, green gram, vegetables etc. It is highly toxic to honeybees, fishes and water inhabiting organisms. It is compatible with most of the plant protection chemicals. It acts by blocking the neuronal sodium channels in larvae causing paralysis and ultimate death. It is formulated as 14.5% SC and marketed under the trade names 'Indoxa Gold', 'K Indox', 'King Carb', 'King Doxa' etc. It is applied at the rate of 200 ml. per acre as spray.

## Growth regulator group

i) **Novaluron ($C_{17}H_9ClF_8N_2O_4$).** Technically it is '(RS) - 1 - (3 - chloro - 4 - 1, 1, 2 - trifluoro - methoxyethoxy) phenyl - 3 - (2, 6 - difluorobenzoyl) urea. It is mainly a stomach insecticide, but has some contact action also. It is capable of controlling a wide range of insects belonging to the orders Lepidoptera, Coleopters, Diptera, Homoptera etc. infesting many crops. *Spodoptera litura* and *Helicoverpa armigera* are also effectively controlled by this insecticide. The insecticide inhibits chitin formation during the larval stages, especially during the early instars, which synthesize more of chitin. This prevents further growth of the larvae resulting in their death. The insecticide has no ovicidal action, but the young larvae that hatch out from the eggs are easily killed. However, it is safe to the beneficial insects. It is formulated as 10% EC and marketed under the trade names 'Pedestal', 'Rimon', 'Remaster', 'Novamax' etc. It is applied at the rate of 400 ml. per acre as spray.

## Sulfite ester group

i) **Propergite ($C_{19}H_{26}O_4S$).** It is technically '2 - (4 - tert - butylphenoxy) - cyclohexyl - prop - 2 - yne - 1 - sulfonate'. It is a systemic, contact insecticide with some amount of fumigant action and belongs to the sulfite ester group of insecticides. It is capable of controlling 36 species of mites such as, red spider mite, pink mite, scarlet mite, purple mite, red mite, two - spotted mite etc. infesting various crops such as, cotton, chilli, tomato, egg plant, okra, soybean, apple, tea, grapes etc. It can control mites, which have developed resistance to other insecticides. It interferes with the key mite enzyme system, which causes interruption of normal metabolism, respiration and electron transport functions in the nervous system resulting in paralysis and death of the mites. It is compatible with most of the commonly used plant protection chemicals. It is highly toxic to fishes and other water inhabiting organisms. It is formulated as 57% EC and marketed under the trade names 'Omite', 'Simbaa', 'Optra', 'Mite Kill', 'Teeka', 'Miti Kill' etc. It is applied at the rate of 400 ml. per acre as spray.

## Azomethine group

i) **Pymetrozine ($C_{10}H_{11}N_5O$)**. It is technically '(E) - 4, 5 - dihydro - 6 - methyl - 4 - (3 pyridylmethyleneamino) - 1, 2, 4 - triazin - 3 - (2 H)' It is a contact, systemic, translaminar insecticide, effective against sucking pests such as aphids, white flies, plant hoppers, pollen beetles etc. infesting crops such as, rice, cotton, potato, vegetables, tobacco, fruit crops, hop, ornamentals, mango etc. It is especially recommended for the control of rice brown plant hopper and mango hopper. It is effective in controlling the pests at the juvenile and adult stages. It acts as a feeding deterrent by affecting the nerves controlling the feeding system and causes paralysis of the hind legs, as a result the insects fall down to the ground and die of starvation. It is formulated as 50% WG and marketed under the trade names 'Chess', 'Pymax', 'Apply', 'Inbox', 'Suruga', 'BPH Super', 'Carom', 'Ju-claim', 'Metapro' etc. It is applied at the rate of 150 gm. per acre as spray.

## Spinosyn group

i) **Spinosad**. It is a mixture of two naturally occurring substances *viz.*, **Spinosyn A ($C_{41}H_{65}NO_{10}$)** and **Spinosyn D ($C_{42}H_{67}NO_{10}$)** having insecticidal property. It is synthesized by a soil inhabiting Actinomycete - *Saccharopolyspora spinosa.* It is primarily a contact insecticide, but acts as an ingestion poison also. It is effective in controlling a wide variety of insect pests such as, aphids, thrips, jassids, hoppers, shoot and fruit borer, pod borer, leaf miner, girdle beetle, semi looper, American cotton boll worm, fruit flies, diamond - back moth, mosquitoes, ants, spider mite etc. infesting several crops such as, field crops, cotton, chilli, vegetables, egg plant, red gram, horticulture crops etc. It is safe to natural enemies of insect pests. It is a nicotinic acetyl choline receptor that acts on the nervous system developing hyper excitation and causes their muscles to flex uncontrollably leading to paralysis and ultimate death of the insects within one or two days. It is quite safe to natural enemies of insect pests. It is formulated as 2.5% EC and marketed under the trade names 'Success', 'Spino 25', 'Spinosad 2.5' etc. and as 45% SC and marketed under the trade names 'Taffin', 'Natroba', 'Tracer', 'Spinter', 'Conserve' etc. 2.5% EC is applied at the rate of 300 ml. per acre as spray and 45% SC is applied at the rate of 80 ml. per acre as spray.

## Tiptronic acid group

i) **Spiromesifen ($C_{22}H_{30}O_4$).** Technically it is '2 - oxo-3 - (2, 4, 6 - trimethylphenyl) - 1 - oxaspiro (4, 4) - non - 3-en - 4 yl - 3, 3 - dimethylbutanoate'. It is primarily an acaricide and contact, non - systemic insecticide effective against all kinds of mites such as, red spider mite, purple mite, scarlet mite etc. and white flies infesting several crops such as cotton, chilli, brinjal, vegetables, fruit trees, apple, tea etc. It affects all growth stages of the pests from eggs, nymphs and adults. It acts in an unique way by inhibiting lipid biosynthesis especially triglycerides and free fatty acids and kills the insects. It is highly toxic to all aquatic life. It is compatible with most of the commonly used plant protection chemicals. It is formulated as 22.9% SC and marketed under the trade names 'Danfuran', 'Chemifider', 'Oberon', 'Offmite', 'Voltage' etc. It is applied at the rate of 200 ml. per acre as spray.

List of pesticides / formulations banned in India as on 01 - 10 - 2022 is given in **Appendix - 8**

List of pesticides / formulations approved by the Government of India for plant protection purposes is given in **Appendix - 9**

## Mixed insecticidal formulations

According to available evidences, insects came into existence in this planet Earth several millions of years before man appeared on this Earth. To satisfy his basic needs of food, clothing and shelter, man started growing different crops. In the meanwhile insects had gradually diverted their original feeding habits and started feeding and destroying the crops raised by man. Under these circumstances, man considered such insects as his prime enemy and tried to destroy them by adopting various means. From time immemorial man had used plants and products obtained from plant origin as well as those from animal origin for the control of harmful insects. Side by side, some of the inorganic chemical compounds were also used for this purpose. After the invention of the first synthetic organic phosphorus insecticide, 'Dichloro diphenyl trichloro ethane' (DDT) in 1939, several other organo phosphorus, organo chlorine, carbamate, pyrethroid and many other pesticides were introduced for the control of insect pests. These pesticides were found to be non - specific, highly toxic and the residual toxicity persisted in the applied parts for a considerably long time. Further most of these pesticides were not compatible with other plant protection chemicals. Because of their high mammalian toxicity many of these supposed to be old generation pesticides were banned by the Government for plant protection use. Subsequently several new pesticides have been introduced for pest control. These new generation pesticides, which are more specific and recommended for the control of certain specific insect pests are capable of controlling many other pests simultaneously. besides the target pests. Though they are highly toxic to insect pests, they have less mammalian toxicity and low residual toxicity. They are mostly compatible with other pesticides, fungicides and fertilizers. They have quick, knock - down effect.

Of late, many mixed insecticide formulations have been introduced. These mixed insecticide formulations can control more number of insect species than the number of insect species that can be controlled by any one of the component insecticide. When one of the insecticide can control some of the insect species, the other component insecticide may be able to control other species of insects. Mixed insecticide and fungicide formulations can control specific insect pests and fungal pathogens simultaneously.

i) **Acephate + Imidachloprid**. In this mixed formulation Acephate 50% is mixed with Imidachloprid 1.8%. This mixture is a broad spectrum, systemic, contact and stomach insecticide and is more effective in controlling more species of insect pests. It can control many sucking and chewing type of insect pests such as, aphids, thrips, white flies, mealy bugs, green leaf hopper, brown plant hopper, sugarcane hopper, mango hopper, stem borer, leaf miner, leaf folder, fruit borer, boll worms, army worm, weevils etc., infesting various crops such as, rice, cotton, cereals, maize, sugar beet, sugarcane, tomato, potato, vegetables, citrus plants, cucurbits, tobacco, fruit trees etc. It has got quick, knock - down effect. It is marketed under the trade names 'Lancer Gold', 'Fighter Gold' etc. It is applied at the rate of 400 g. per acre as spray.

ii) **Chlorantraniliprole + Lambda Cyhalothrin**. In this mixed formulation Chlorantraniliprole 9.3% is mixed with Lambda Cyhalothrin 4.6%. This mixture is a broad spectrum, non - systemic, contact and stomach insecticide and is more effective in controlling more species of insect pests than the individual component insecticide. It has got ovicidal, larvicidal and adulticidal properties and can control a wide range of sucking and chewing insect pests such as, Lepidopterans, boll worms, pod borers, stem borer, leaf folder, leaf miner, green leaf hopper, stem fly, girdle beetle, semilooper, beetles, thrips, spider mites, diamond - back moth etc. infesting major crops such as, rice, cotton, sugarcane, potato, tomato, tobacco, chilli, soy. bean, okra, pigeon pea etc. It has got quick knock - down effect. It is marketed under the trade name 'Ampligo' and is applied at the rate of 100 ml. per acre as spray.

iii) **Chlorpyriphos + Alphacypermethrin**. In this mixed formulation Chlorpyriphos 16% is mixed with Alphamethrin 1.0%. This mixture is a broad spectrum, non - systemic contact, stomach insecticide with some amount of respiratory action and is more effective in controlling more species of insect pests than the individual component insecticide. It can control a wide range of sucking and chewing insects belonging to the Orders Lepidoptera, Coleoptera, Diptera and Hemiptera infesting over 100 crops, including rice, cotton, chilli, groundnut, cereals, sorghum, maize, fruit crops, vegetables, potato, tobacco, beet, soybean, sunflower etc. It is marketed under the trade names 'AC-116', 'Chilorthrin', 'Toofan', 'Raffle', 'Alphasulphan' etc. and is applied at the rate of 500 ml. per acre as spray.

iv) **Chlorpyriphos + Cypermethrin**. In this mixed formulation, Chlorpyriphos 50% is mixed with Cypermethrin 5%. This mixture is a broad spectrum, non - systemic, contact and stomach poison with respiratory action and is more effective in controlling more species of pests than the individual component insecticide. It can control a wide range of sucking and chewing insects pests such as, aphids, thrips, white flies, mealy bugs, scale insects, jassids, leaf hopper, plant hopper, stem borer, leaf folder, leaf miner, case worm, army worm, bud worm, cutworm, pod borer, fruit borer, grasshopper etc. infesting several crops such as, rice, cotton, chilli, groundnut, soybean, sunflower, tobacco, legumes, vegetables, fruit trees, vines etc. It is marketed under the trade names 'Yuja', 'Bombard', 'Double Star', 'Fighter', 'Super D', 'Combi King', ' Koranda 505' etc. It is applied at the rate of 400 ml. per acre as spray.

v) **Chlorpyriphos + Fenobucarb**. It is a combination of Chlorpyriphos - 21 % + Fenobucarb - 10.5 %. It is a broad spectrum, non - systemic, contact and stomach insecticide with some vapor action also. It is capable of controlling a wide range of flying pests, chewing and boring insect pests, bugs, jassids, hoppers, thrips, white flies etc infesting several crops such as, rice, corn, cotton, tomato, tobacco, cucurbits, onion, legumes etc. It is marketed under the trade name 'Perfek' and it is applied at the rate of 300 ml. per acre as spray.

vi) **Quinalphos + Cypermethrin**. In this mixed formulation Quinalphos 20% is mixed with Cypermethrin 3%. This mixture is a contact and stomach poison with translaminar systemic activity. It is mainly used for the control of Lepidopterous pests such as, American boll worm, spotted boll worms, army worm, shoot borer, fruit borer, pod borers, jassids etc. infesting cotton, brinjal, vegetables etc. It is marketed under the trade names 'Viraat', 'Cyqueen', 'Kwin Cyp', 'Victory' etc. It is applied at the rate of 250 ml. per acre as spray.

vii) **Ethion + Cypermethrin**. In this mixed formulation Ethion 40% is mixed with Cypermethrin 5.0%. This mixture is a broad spectrum, non - systemic, contact and stomach poison effective against both sucking and chewing insect pests. It is more effective in controlling more species of pests effectively than the individual component insecticide. It has got ovicidal, larvicidal and adulticidal property. It can control a wide range of insect pests such as, aphids, thrips, white flies, hoppers, caterpillars, beetles and mites infesting vegetables, pulses and all major crops, especially borer pests of cotton, brinjal, okra etc. It is marketed under the trade names 'Ethiccyper', 'Nagata', 'Hero No.1', 'Colfos', 'Cyperthion', 'Cobra 404' etc. It is applied at the rate of 400 ml. per acre as spray.

viii) **Imidachloprid + Beta - cyfluthrin**. In this mixed formulation Imidachloprid - 19.81 % is mixed with Beta - Cyfluthrin - 8.49 %. This mixture is a broad spectrum, systemic, contact and stomach poison. It can control a wide range of sucking and chewing insect pests such as, aphids, thrips, white flies, hoppers, caterpillars, beetles and mites infesting various crops such as, Cotton, turf, ornamentals, hops, cereals, cotton, potato, chillies, tomato, fruit trees etc. It is also used for the control of household insects such as, cockroaches, bed bugs, silver fishes etc. It is highly toxic to fishes, water inhabiting organisms and honeybees It is marketed under the trade name 'Solomon'. It is applied at the rate of 100 - 125 ml. per acre as spray.

ix) **Imidachloprid + Ethiprole**. In this mixed formulation Imidachloprid 40 % is mixed with Ethiprole - 40 %. It is a systemic, contact and stomach insecticide and is highly effective in controlling sucking pests such as aphids, thrips, white flies, hoppers and caterpillars . It is very effective in controlling brown plant hopper, green leaf hopper and white backed plant hopper infesting rice crop and prevents hopper burn in the later crop stages. It is marketed under the trade name 'Glamore' and is applied at the rate of 50 - 60 g. per acre as spray.

x) **Deltamethrin + Buprofezin**. In this mixture Deltamethrin - 0.72 % is mixed with Buprofezin - 5.65 %. It is a non - systemic, contact and stomach poison with persistent toxicity and can control sucking and chewing insect pests, as well as mites. It is recommended for the control of brown plant hopper and leaf folder and mealy bugs of rice. It is effective against pests, which have developed resistance against other pesticides. It is also used for the control of pests infesting cotton, chilli, brinjal etc. It has got ovicidal, larvicidal and adulticidal properties.

It inhibits moulting of larvae and nymphs leading to their death. It suppresses oviposition by adults and the treated adults lay sterile eggs. It is marketed under the trade names 'Black Apollo', 'Buprothrin', 'Sydney', 'Tusker', 'Devikey' etc. It is applied at the rate of 500 ml. per acre as spray.

xi) **Deltamethrin + Triazophos**. In this mixture Deltamethrin - 1.0 % is mixed with Triazophos 35 %. It is a non - systemic, broad spectrum insecticide and acaricide having contact and stomach action. It is effective against sucking and chewing types of pests. It can control a wide range of pests such as, cotton spotted boll worm, pink boll worm, American boll worm, white flies, shoot and fruit borer, jassids, aphids, thrips, mites, lady bird beetles etc. infesting various crops such as rice, cotton, brinjal, soybean, chilli etc. It is marketed under the trade names 'Sparker', 'Delshot', 'Magic', 'Fulstop - D', 'Anaconda Plus', 'Fusion', 'Tricada', 'Scoop' etc. It is applied at the rate of 400 ml. per acre as spray

xii) **Imidachloprid + Fipronil**. In this mixture Imidachloprid - 40% is mixed with Fipronil - 40 %. It is a broad spectrum, systemic, contact and stomach poison. It is effective in controlling a wide range of sucking and chewing insect pests such as, aphids, thrips, white flies, jassids, hoppers, stem borers white grubs, caterpillars etc. infesting several crops such as, sugarcane, groundnut, rice, field and horticultural crops. It is marketed under the trade names 'Velma 80', 'Gharada Police', 'Nashak', 'Shirasagi', 'End Task', 'Panther', 'Umpire', 'Lessenta' etc. It is applied at the rate of 100 - 120 g. per acre as spray.

xiii) **Flubendiamide + Thiachloprid**. In this mixture Flubendiamide - 19.92 % is mixed with Fipronil - 19.2 %. It is a broad spectrum, systemic, contact and stomach poison and can control a wide range of chewing insect and some sucking insect pests such as yellow stem borer, leaf folder, fruit borer, thrips, aphids, white flies etc. infesting crops such as, rice, chilli, cotton, cereals, tobacco, tomato etc. It is marketed under the trade names 'Belt Expert', 'Prayan', 'Agro Star Magna' etc. It is applied at the rate of 100 ml. per acre as spray.

xiv) **Indoxacarb + Acetamiprid**. In this mixture Indoxacarb - 14.5 % is mixed with Acetamiprid - 7.7 %. It is a broad spectrum, systemic, contact and stomach poison and can control a wide range of sucking and chewing insect pests such as, jassids, white flies, hoppers, thrips, boll worms, fruit borers and other Lepidopterous insect pests infesting various crops such as, cotton, chliies, rice, cereals etc. It is marketed under the trade names 'Kito', 'Kara Plus', 'Indoprid', 'Caesar' etc. It is applied at the rate of 200 ml. per acre as spray.

xv) **Profenophos + Cypermethrin**. In this mixture Profenophos - 40 % is mixed with Cypermethrin - 4 %. It is a broad spectrum, non - systemic,contact and stomach insecticide with translaminar action. It is effective in controlling a wide range of sucking and chewing pests infesting a number of food and horticultural crops such as rice, sugarcane, mango, potato, chilli etc. It is especially recommended for the control of cotton pests. It is marketed under the trade names 'Roket', 'Cyprofen', 'Cypermen', 'Hinawari', 'Propcyp', 'Prolitrin', 'Ajanta Super' etc. It is applied at the rate of 400 ml. per acre as spray.

xvi) **Thiamethoxam + Chlorantraniliprole**. In this mixyure Thiamethoxam - 1 % dust is mixed with Chlorantraniliprole - 0.5 % granules. It is a broad spectrum, systemic insecticide capable of controlling many chewing insect pests such as, leaf folder, stem borer, early shoot borer and sucking insect pests such as, aphids, jassids, hoppers, thrips,, white flies etc. infecting many crops such as rice, sugarcane, corn, chilies' etc. It is marketed under the trade names 'Vasishta', 'Virtako' etc. It is applied at the rate of 2.5 kg. per acre. The mixture is mixed with 10 kg. of sand and broadcasted over the field when the soil is moist. The insecticide is quickly absorbed by the roots of the germinating plants and translocated all over the tissues.

xvii) **Thiamethoxam + Lambda Cyhalothrin.** In this mixture Thiametoxam - 12.6 % is mixed with Lambda Cyhalothrin - 9.5 %. It is a broad spectrum, systemic, contact and stomach poison and is capable of controlling a wide range of sucking and chewing insect pests such as, Jassids, thrips, white flies, leaf hoppers, shoot fly, boll worms, aphids, leaf eating caterpillars, stem borer, fruit borer, tea mosquito bugs, semi loopers, girdle beetle etc. infesting several crops such as, rice, cotton, chilli, maize, groundnut, tomato, tobacco, soybean, tea etc. It is marketed under the trade names 'Alika', 'Lruka - MC', 'Chakravarti', 'Long Star' etc. It is applied at the rate of 80 ml. per acre as spray.

xviii) **Thiamethoxam + Chlorantraniliprole**. In this mixture thiamethoxam - 17.5 % is mixed with Chlorantraniliprole - 8.8 %. It is a wide spectrum, systemic, contact and stomach poison and is capable of controlling many sucking and chewing insect pests such as, Lepidoperous caterpillars, aphids, jassids, thrips, white flies, hoppers, white flies etc., infesting cotton, fruit plants, citrus plants, crop plants, tobacco etc. It is marketed under the trade names 'Volium Flexi', 'Aandhi' etc. It is applied at the rate of 100 ml. per acre as spray

**Mixed insecticide - fungicide formulations**

i) **Flubendiamide + Hexaconazole**. In this mixture Flubendiamide - 3.5 % is mixed with Hexaconazole - 5 %. It is a broad spectrum, systemic and contact pesticide primarily recommended for the control of rice stem borer and leaf folder as well as sheath blight of rice. However, it is effective in controlling a wide range of Lepidopterous pests infesting rice, chilli, tomato, brinjal, cabbage, black gram, soybean etc. besides *Helicoverpa-, armigera* and *Spodoptera litura*, It is also effective in controlling many fungal diseases such as rusts, leaf spots and powdery mildew of mango, grapes, black gram etc. It is marketed under the trade names 'Origin', 'Carter' etc. It is applied at the rate of 400 g. per acre as spray

**Fumigants**. Fumigants are gaseous poisons used to destroy insect pests, nematodes, rodents and other such pests. Fumigants are more effective under airtight conditions. They enter into the body of larvae, pupae and grown - up adults through the respiratory tracts, damage the tissues by different ways and cause their death. Fumigants can destroy even the eggs by entering inside through the chorion and bring about their death before hatching. Pests inhabiting deep burrows, cracks, crevices and other such hiding places, which cannot be killed by the application of sprays, dusts etc., can be effectively killed by fumigants as they can easily penetrate into such hiding places. Fumigants can perform better when the temperature is fairly high (20° - 25°C). It is preferable to take up fumigation in darkness or during the night time, Fumigants are widely used for killing various kinds of storage pests, soil - inhabiting insect pests, as well as soil nematodes. They are highly poisonous and hence should be used very carefully under expert supervision. To facilitate exit of the residual toxic fumes after completion of fumigation, provision should be made for good aeration.

i) **Hydrogen cyanide** or **Hydrocyanic acid (HCH)**. It is one of the most commonly used fumigant and is widely used to eliminate pests of stored grains, pests of other agricultural products, rats etc. It is volatile and highly inflammable. It is available as dust or granular formulation under the trade names 'Cymag', 'Cyanogas' etc. Cyanogas contains 25% of Hydrocyanic acid and Cymag contains 20% of Hydrocyanic acid. In godowns, warehouses etc. Cyanogas is used at the rate of 1.5 oz. (42 g.) per every 1,000 cu. ft. volume. The treatment should be continued for a period of 24 hours and should be done preferably under darkness and under air tight conditions. For rat control, Cyanogas pump is used to pump in Cyanogas dust into the rat burrows at 6.0 - 10.0 g, per rat burrow under pressure.

ii) **Carbondisulphide**. It is a highly inflammable substance and hence it is not used as such, but used in combination with carbon tetrachloride in the ratio 1 : 4. It is available commercially in liquid form and it volatilizes very easily. It is used for the control of storage pests in godowns, ware houses, flour mills etc. It has good penetrating power and does not leave any toxic residues. If it is used to control pests in moist or not well- dried seeds, the germination of the seeds is adversely affected. The treatment should be continued for 36 - 72 hours for good results. To treat 1,000 kg. of grains, a quantity of 3.5 - 5.0 lit. of the fumigant is required or for every 1,000 cu. ft. volume, 4.5 lit. of the fumigant should be used.

iii) **Chloropicrin** or **Trichloronitromethane**. It is available in a liquid form under the trade names 'Picfume', 'Acquinite' etc. It is not easily inflammable and has got good penetrating quality. It is also used as tear gas. It is a low volatile compound and the gaseous residue persists for a longer time and cannot be removed easily. Because of its injurious effect, it should not be used for fumigating germinating seeds or growing plants. It is effective in controlling pests found in flour mills, godowns etc. It is applied at the rate of 450 - 1,350 ml. for every 1,000 cu. ft volume as spray or may be sprinkled on the bags packed with seeds. It can also be used for the control of soil - inhabiting nematodes.

iv) **Ethylene dichloride - carbon tetrachloride (EDCT)**. It is a mixture of Ethylene dichloride - 3 parts and Carbon tetrachloride - 1 part. It is one of the most commonly used fumigant for the control of storage pests. The fumigant, which is available in a liquid form is not inflammable. It is very effective at moderate temperatures of about 24°C and can be made to volatilize by placing the liquid in wide mouthed vessels in several places or by spraying. It is used in godowns, ware houses etc. at the rate of 4.5 - 8.0 lit. for every 1,000 cu. ft. of volume and the treatment should be continued for a period of 24 hours.

v) **Dichloropropane - dichloropropene (DD mixture)**. It is available under the trade name 'Telone' as a liquid formulation. It can effectively control all kinds of nematodes and other soil - inhabiting pests. The residual toxicity of this fumigant may persist in the soil for many days after treatment and may cause injury to the next crop. Hence it should be injected into the soil 2 - 3 weeks prior to planting at a depth of 18 - 20 cm. and at a spacing of 25 - 30 cm. with a soil injecting gun at the rate of 90 - 115 lit. per acre. Before planting, the soil is thoroughly stirred to let out the residual fumes in the soil. It taints potato tubers and carrots grown in treated soils.

vi) **Dichloropropane**. It is available under the trade names 'Nemagon', 'Fumazone' etc. in a liquid form. It can effectively control many of the soil - inhabiting nematode species. It can be applied either before or after planting. It is more effective when the temperature is more than 20°C. It can be mixed with water and sprayed on the soil surface before planting or mixed with irrigation water. Depending upon the growth stage of the crop, it is applied at the rate of 4.0 - 24.0 lit. per acre. It is injurious to potato and tobacco crops.

vii) **Paradichlorobenzene**. It is a white crystalline chemical substance that volatilizes slowly into a non - inflammable gas. It has a characteristic ether - like odor. It is recommended for the control of carpet beetle, clothes moth, insects damaging museum specimens etc. at a dosage of 12 g. / $m^3$ (340 g. / 100 cu. ft. of volume). It is not used for treating grains meant for either seed or for human or livestock consumption, as it adversely affects the germination of seeds and imparts a highly objectionable odor and taste.

viii) **Naphthalene**. It is a coal tar derivative, long used as a fumigant. It is a white crystalline material available in the form of small balls. It vaporizes slowly at normal temperatures. The gas is inflammable when mixed with air. It is widely used to control insect pests infesting and damaging woolen materials, silk and cotton clothing, blankets, carpets, doormats, museum specimens etc. It is also used in green houses for the control of insect pests. It is used at the rate of 1.0 g. / $m^3$ (28 g. / 100 cu. ft. of volume). It is not used for treating seeds or grains meant for consumption.

ix) **Vapam**. It is a nematicide available under the trade name 'Vapam' as a liquid formulation. Its residual toxicity persists in the soil for several days after application. Hence it should be injected into the soil one or two weeks prior to planting at a depth of 18 - 20 cm. and at a spacing of 25 - 30 cm. with a soil injecting gun at the rate of 90 lit. / acre. Before planting the soil is thoroughly stirred to let out the residual fumes in the soil.

## Systemic insecticides

Systemic insecticides are chemical compounds, which are absorbed by the plant parts and translocated to all the parts of the plant tissues through the vascular system along with the sap, thus remain inside the plant tissues and provide temporary immunity to the plants from pest attack. The pests, which feed on the treated plants ingest the insecticide along with the sap and are killed. Most of the systemic insecticides are organic stomach poisons, while some of them act as contact poisons also. These compounds are found among 'Organophosphates', 'Organochlorine compounds', 'Carbamates', Neonicotinoids' and some other groups of insecticides. They are very effective against sucking pests. The vast majority of systemic insecticides move upward in the plant tissues (Acropetal), concentrate in the leaf tissues and then disintegrate as a result of oxidative and hydrolytic metabolism. Only very few of them are translocated downward and move toward the root portion. Most of these insecticides have fairly long residual action. Though some of the insecticides such as, malathion, diazinon, nicotine, lindane etc. move along the plant tissues to some extent, their movement is only translaminar and are not translocated throughout the plant tissues as in the case of true systemic insecticides.

Systemic insecticides are absorbed through the roots, stomata, cuticle, axillary buds, epidermis, germ pores of seeds, injuries in the cuticle etc. and then translocated throughout the plant tissues. Systemic insecticides are formulated as emulsifiable concentrates, water - soluble liquids, granules etc. They can be applied in several ways such as, seed dressing, foliar spraying, broadcasting in the fields, spot application, furrow application, whorl application, root feeding, stem injection etc. Systemic insecticides can be used effectively for the control of insect and non - insect vectors such as, leaf and plant hoppers, aphids, white flies, thrips, nematodes etc., which transmit many virus and mycoplasma diseases. Certain amount of moisture is required for effective functioning of systemic insecticides and absorption of the toxicant is greater at fairly higher temperatures through younger leaves.

### Advantages of using systemic insecticides

i) Systemic insecticides are absorbed rapidly into the plant system and hence the atmospheric conditions do not affect them much and as such their residual toxicity persists in the plant tissues for a longer period and afford long lasting protection against pests.

ii) Most of the systemic insecticides do not harm beneficial insects such as, parasitoids, predators, pollinators etc.

iii) Pests do not develop resistance against systemic insecticides easily

iv) Even when the foliage is much less, systemic insecticides can enter into the plant system easily through various natural openings in the plant body or through injuries and prove to be effective in controlling the pests.

v) The chances of these insecticides causing any kind of direct or indirect injurious effects in the treated plants is very much less.

vi) The after effects due to residual toxicity of these insecticides is also very much less.

## Disadvantages of using systemic insecticides

i) Only after sufficient quantity of insecticide is ingested by the insects to reach the lethal dose, they are killed. So, it may take a longer period for the pest to be destroyed.

ii) Some of these insecticides are highly toxic to humans and livestock

iii) Systemic insecticides are more specific in destroying only certain pest species.

iv) Few systemic insecticides, though capable of destroying different types of pests, including sucking pests, repeated use of these insecticides may induce resistance in some species of sucking pests such as, aphids, white flies, plant hoppers etc.

## Resistance to insecticides

When certain insecticides are used repeatedly over a short period of time, few insect species may develop sub - species or form species, which may develop resistance against these insecticides. Such sub - species or form species can withstand, survive and reproduce even when these compounds are applied at concentrations, which can destroy other insects of the same species, at the same time the future generations of such resistant insects may also develop resistance. These form species cannot be destroyed even by the application of much higher doses of such chemicals. Insecticide resistance may develop genetically or by mutation or may be due to development of certain physiological or morphological changes in the pests. Some strains of insects may show a natural tolerance to a particular pesticide possibly due to its ability to detoxify it by enzyme activity. If such a strain constitutes a sizable level of the total population, there is every possibility that this population may survive that pesticide treatment and go on to multiply unabated. A large proportion of offsprings of the pest will also be resistant to that particular pesticide. Resistance is likely to develop rapidly in species, which have numerous generations within a short span of time.

**Melander** (1914) was the first Scientist to establish that the San Jose scale insect had developed resistance against lime - sulphur. Later **Quayle** (1916) found out that the California hard scale insect on citrus had developed resistance against Hydrogen sulphide fume. Subsequently till 1945, 13 insect species including few ticks were found to have developed resistance to Arsenic insecticides, Hydrogen cyanide, Selenium, Rotenone etc. In 1847, resistance of some flies against DDT was reported from countries such as, Italy and Sweden. Up to 1960, about 137 insect species were reported to have developed resistance against insecticides belonging to organochlorine, organophosphorus and carbamate groups. Recently it was found that the rice green leaf hopper had developed resistance against Diazinon. In India, species of mosquitoes, which are carriers of some of the most serious diseases of man have been reported to have developed resistance against DDT, HCH, Dieldrin, organophosphorus insecticides etc.

Resistance of some pests against pesticides may be due to several mechanisms. The most important mechanism to organophosphorus compounds is physiological resistance. This involves the genetically controlled process of enzymatic hydrolysis and detoxification of the pesticide in the insect system. Another mechanism is behavioristic resistance in which the pest species avoid doses of toxic substances, which would otherwise be lethal. Mosquitoes avoid surfaces treated with DDT.

Several public health, household, veterinary, crop and stored grain pests have developed resistance against specific pesticides in India. Mosquito - *Culex pipens fatigans* transmitting malaria and other vector diseases in man is resistant to DDT and HCH. The bed bug - *Cimex lectularis* is resistant to DDT, HCH and a few organophosphorus compounds. The human body louse *Pediculus humanus corporis* is resistant to HCH. The rat flea - *Xenopsylla cheopis* is resistant to DDT. The house fly - *Musa domestica* is resistant to DDT and HCH.

Singhara beetle - *Galarucella birmanica* is resistant to DDT and HCH. Tobacco caterpillar or cut worm - *Spodoptera litura* is resistant to HCH, Malathion, Endosulfan and Carbaryl. Resistance to parathion, DDT, Fenitrothion, Cypermethrin, Fenvalerate, Deltamethrin, Quinalphos and Endosulfan has been reported in cabbage diamond - back moth - *Plutella xylostella.* Tolerance to Malathion and Dimethoate has also been noticed. American cotton boll worm *Helicoverpa armigera* has also developed resistance to synthetic pyrethroids such as, Cypermethrin and Fenvalerate.

The flour beetle - *Tribolium castaneum* is resistant to DDT, Malathion, Lindane and Phosphine. Further resistance to Malathion, Lindane and Phosphine in the rice weevil - *Sitophilus oryzae,* to Malathion and Lindane in the grain beetle *Oryzaephilus surinamensis,* to Phosphine in khapra beetle - *Dermestes granarium,* to leather beetle - *Dermestes maculatus* and other instances have also been reported.

## Insecticide resistance management

Insecticide resistance problem has become a very serious issue during the recent past and several measures have been suggested to overcome this problem.

Judicious application of pesticides, only when it is absolutely necessary, selection of suitable pesticides for which the target pest has not developed resistance, application of pesticides at the most vulnerable growth stage of the pests, using synergists, which will enhance the toxicity of a particular pesticide by inhibiting the detoxification mechanism in the pest, using correct dose of pesticides as per recommendations, alteration of chemicals with unrelated modes of action, using formulations containing two or more compounds with different modes of action, application of an organophosphorus pesticide following use of synthetic pyrethroids once or twice, avoiding use of the same pesticide repeatedly over a short period of time, adopting various sowing times to avoid peak risk periods, use of pest resistant varieties of crops, adopting crop rotation practice to avoid build - up of the pest species, adopting strict crop hygiene methods, adopting biological control measures and such other strategies will go a long way in not only controlling the pests effectively but also prevent development of resistance in insects to pesticides.

## Preparation of insecticidal spray fluids

Most of the insecticides are not formulated with 100 per cent active ingredient, but are dissolved in suitable solvents, so that they can disperse or dissolve in water easily and also to prolong their shelf life. Along with the solvents, emulsifiers, spreaders and stickers are also added to increase their spreading and sticking capacity, and formulated with a specific concentration of the active ingredient. For example in chlorpyriphos 20EC, only 20% is chlorpyriphos active, while the rest 80% consists of solvents, emulsifiers, spreaders and stickers. Similarly the wettable powder formulations also contain a specific quantity of active ingredient and the rest emulsifiers, spreaders and stickers.

The commercially available or proprietary insecticidal formulation containing a specific percentage of active ingredient is mixed with a specific quantity of water to prepare a spray fluid with a desired percentage of active ingredient.

When different types of spray equipments are used for spraying, the quantity of spray fluid required to cover an unit area may vary. When a high volume sprayer is used to spray small crops such as, rice, finger millet etc., a quantity of 200 litres of spray fluid may be required to cover one acre. On the other hand, when a low volume sprayer is used, a lesser quantity of about 60 litres of spray fluid may be sufficient to cover the same area. However, the quantity of the insecticide formulation to be used in both the cases should be the same. Even if a low volume sprayer is to be used, the quantity of insecticide formulation to be applied should be calculated based on the water requirement for a high volume sprayer and should not be based on the water requirement for the low volume sprayer.

The quantity of spray fluid that may be required to spray one acre of crop by different types of sprayers is furnished in **Appendix - 11**

All spray formulations are recommended to be used at a specific concentration of the active ingredient in the spray fluid to effect maximum control of the pests at the same time to avoid injury to the plants sprayed. To calculate the quantity of pesticide formulation required to prepare a spray fluid of desired concentration of the active ingredient, the following formula is applied:

$$\text{Quantity of the insecticide formulation required to prepare the required quantity of spray fluid of desired strength} = \frac{\text{Quantity of the spray fluid required} \times \text{Concentration of the spray fluid in percentage}}{\text{Percentage of active ingredient in the insecticide formulation}}$$

For example, if 200 litres of spray fluid of 0.035% active ingredient is to be prepared from Endosulfan 35 EC to spray one acre of crop, the above formula can be applied as follows :-

$$\text{Quantity of Endosulfan 35 EC required} = \frac{200 \text{ X } 0.035}{35} = \frac{7}{35}$$

$$= 0.2 \text{ litre or } 200 \text{ ml.}$$

In 200 litres of water, 200 ml. of endosulfan 35 EC is mixed to obtain a spray fluid of 0.035 % concentration.

When a low volume sprayer is used, which may require a lesser quantity of only about 60 litres of spray fluid to cover the same area of one acre, 200 ml. of endosulfan 35 EC should be mixed with 60 litres of water. The quantity of the formulation required should not be calculated based on the lesser quantity of 60 litres of water.

Similarly, if 300 litres of spray fluid of 0.1% active ingredient is to be prepared from carbaryl 50 WP, the same formula can be applied thus ;

$$\text{Quantity of Carbaryl 50 WP required} = \frac{300 \text{ x } 0.1}{50} = \frac{3}{5}$$

$$= 0.6 \text{ kg. or } 600 \text{ g.}$$

When 600 g. of carbaryl 50 WP is mixed with 300 litres of water, a spray fluid with 0.1% concentration is obtained.

Likewise, if a specific quantity of an insecticide formulation is mixed with a specific quantity of water to prepare a spray fluid, the active ingredient present in the spray fluid in percentage can be calculated by using the formula given below :

$$\text{Percentage of active ingredient present in the spray fluid} = \frac{\text{Quantity of the insecticide formulation used} \times \text{Percentage of active ingredient in the insecticide formulation}}{\text{Quantity of spray fluid prepared}}$$

For example, if 200 ml. of malathion 50 EC is mixed with 200 litres of water to prepare a spray fluid, the active ingredient present in the spray fluid is calculated as follows :

Quantity of malathion 50 EC used = 200 ml. or 0.2 litre

Percentage of active ingredient in the spray fluid $= \frac{0.2 \times 50}{200} = 0.05\%$

Malathion 50 EC, 200 ml. or 0.2 lit. mixed in 200 litres of water will give a spray fluid of strength 0.05 %

Ready reckoner to find out the quantity of various insecticide formulations required to prepare desired concentration of spray fluid (ml. / g. for every 100 litres of water) is given in **Appendix - 12**

## Precautions in handling pesticides

All pesticides including insecticides, acaricides, nematicides, rodenticides, weedicides and many of the fungicides are toxic not only to insects and other pests, but also to human beings and livestock. Pesticides may enter the body of man and livestock by ingestion through mouth, inhalation by nose or absorption by bare skin and exert their poisonous effect. If such poisonous chemicals are not handled properly or if sufficient care is not taken during application, dangerous consequences may result. In large scale use of pesticides, hazards may arise due to accidental or intentional poisoning, operational hazards at the time of application or post - application hazards due to toxic residues. To avoid such hazards, all precautionary measures should be taken at every stage such as, storing, handling, applying etc.

### 1. In storage

i) All stocks of pesticides such as, insecticides, acaricides, nematicides, rodenticides, weedicides, fungicides, fumigants etc. should be stored securely under lock and key, with the labels on the containers intact and kept away from all types of food, stored products, livestock feeds and fodder. They should be kept out of reach of children, domestic animals and birds.

ii) The containers or bags should not be opened in an enclosed space. The dust formulations should be taken out from bags carefully so as to avoid the dust particles from flying and being inhaled.

iii) All toxic chemicals should be stored in their original containers with the labels intact and arranged in such a way that they can be easily seen. Liquid concentrates should not be transferred from their original containers to other containers with some other labels or with no labels.

iv) Pesticides should not be handled with bare hands. Rubber gloves, aprons or protective clothing should be used while handling pesticides.

v) Cautionary or warning notices such as, 'poison', 'skull and cross bone signs' etc. should be displayed prominently in pesticide store houses. Smoking, chewing of tobacco, or use of open fires in the godowns or warehouses should be totally prohibited.

vi) Some pesticides have to be protected from exposure to direct sunlight so as to prolong their shelf life. Such chemicals should be stored inside the room and away from direct sunlight. Wettable powders should be stored under moisture - proof conditions to prevent them from caking.

vii) Pesticide containers or bottles, gunny bags etc. should be disposed off properly by burning or burying them deep under the soil. They should never be used for storing food products, edible oils, food grains etc.

**2. In the fields**

i) Before applying pesticides, the instructions furnished in the labels should be studied carefully and the instructions should be strictly followed

ii) To open pesticide containers, knives regularly used in the household should never be used. For measuring the pesticides, spoon, cups etc. used in the houses should not be utilized. Separate knives, measuring jars, spoons etc. should be kept for these purposes.

iii) Only experienced, physically and mentally sound persons should be engaged to undertake pesticide application work. While handling or applying pesticides, eating, smoking, chewing etc. should be avoided.

iv) Protective coverings such as, face mask, rubber gloves, gum boots etc. should be used.

v) While spraying or dusting, the wind direction should be noted and the discharge nozzle should always face the direction of the wind and not against the wind. Dusting or spraying should never be carried out against the wind current.

vi) Pesticides should not be dusted or sprayed when the crop is in open bloom, as they may cause danger to honeybees and other pollinators.

vii) Pesticides with long residual toxicity should not be applied to vegetables, fruit trees and fodder crops when they are in the fruit bearing or harvesting stages.

viii) Only recommended and correct dosage of pesticides should be used. Only the required quantity of pesticide solution should be prepared and the same should be applied as soon as possible. Stock solution for the whole day should not be prepared and stored. Suitable rod or stick should be used to stir the pesticide fluid and bare hands should not be used.

ix) Pesticides should never be mixed by the side of wells or near channels or other water sources. Plant protection equipments should not be washed near water sources, as contamination of the water may prove lethal to fishes and other water - inhabiting organisms.

x) Clogged nozzles should not be blown with the mouth to clean them. Fine, thin wire may be used o clean the clogged nozzles.

xi) Care should be taken to avoid the toxic materials from coming into contact with the bare skin or eyes. By accident, if the pesticide spills on the bare skin or falls into the eyes, they may be washed thoroughly with fresh water.

xii) While undertaking fumigation work, gas mask should always be used.

xiii) Grains treated with pesticides should not be used for human consumption or fed to cattle, poultry etc. even after washing them.

xiv) Direct mixing of insecticidal dusts with grains to protect them against pest infestation should be avoided.

xv) While spraying or dusting, drifting of the chemicals to nearby water sources, neighboring fields, cattle yards, poultry units etc. should be avoided. When there is strong wind current, spraying or dusting should be avoided.

xvi) Dusting or spraying should be avoided when there is large scale activity of beneficial insects, especially honeybees.

xvii) During pesticide application, if headache, tiredness, vomiting sensation or other such sick symptoms occur, the operation should be discontinued immediately.

### 3. After completion of the work

i) After the days work is over, all the protective clothing, gloves etc. are removed and the hands, feet, face and other uncovered parts of the body should be thoroughly washed with soap and water or a full bath should be taken.

ii) Grazing of cattle or poultry birds and other animals in the pesticide applied fields or nearby areas should be prohibited.

iii) The plant protection equipments should be washed and cleaned in a separate place away from the water sources and kept securely.

iv) First aid should always be kept near and in case of emergency a qualified physician should be contacted immediately.

## Hazards in the use of pesticides

### 1. Directly affecting man and other living organisms

**i) Affecting man while manufacturing and packing of pesticides**. The toxic materials in pesticides may enter the body of man working in pesticide factories at the time of manufacture and packing into various types of containers through the skin, through the nose along with the air while breathing or accidentally through the mouth and may cause many kinds of harmful aftereffects. The toxins thus entering the body may not disintegrate into simpler and harmless substances, but may get accumulated in the tissues and cause dangerous consequences at a later date. These are called **'cumulative poisons'**. Aldrin, dieldrin and other such organochlorine compounds can get accumulated in the body tissues. When large quantities of these poisons enter into the body accidentally, they may prove to be fatal.

**ii) Affecting man while handling and during application of pesticides**. At the time of opening the pesticide containers, while mixing the pesticide with water and at the time of application by various means such as, spraying, dusting, soil application, fumigating etc. the toxic substances may enter the body through the bare skin, through the mouth or through the nose while breathing and may cause serious ill effects.

**iii) Affecting man by ingestion of contaminated food materials.** Harvested food materials such as, grains, vegetables, fruits, greens etc. may contain toxic residues of the pesticides and when these products are ingested, the poison may also enter into the body and cause serious ill effects. Organophosphates, carbamates and most of the systemic pesticides disintegrate in the tissues of plants within a short period and hence the toxic residues do not remain in the plant parts for long periods. However, in the case of many chlorinated hydrocarbons such as, aldrin, dieldrin, heptachlor, toxaphene etc., the toxic residues remain in the plant parts for several days without loosing their toxicity and when the plant parts are consumed, the toxic substances may also enter the body. It has been established that the milk from cattle contains toxic residues of pesticides. Further, it is known that from the body of a pregnant woman, the toxic substances are passed on to the infant in the mother's womb. It has also been found that toxins of some of the pesticides are found even in the mother's milk and are fed to the child through the milk.

iv) **Affecting man by entry into the body through sources other than food materials**. The toxic materials of some of the pesticides may find their way into underground water sources, running water or other surface water sources and may enter into the body of man, when such contaminated water is used for drinking or cooking purposes. The effluents from pesticide factories contain large quantities of toxic materials and are mostly diverted into canals, rivers, streams or ocean. Such toxic materials may find their way into underground water sources and pollute them. The toxic fumes emanating from pesticide factories may enter the body while breathing the contaminated air. The toxic materials, which may be floating in the air may be brought down along with the rains and may contaminate water sources. When such water is used for drinking, it may cause serious ill effects.

v) **Affecting cattle, poultry and other livestock**. To meet the requirement of food and for his personal needs, man raises domestic animals, poultry birds etc. Toxic materials of pesticides may enter into the body of such livestock through feeds or while feeding on vegetation, which may contain toxic pesticide residues. Pesticides also get access to livestock when they are treated with pesticides to destroy ectoparasites infesting them or when the cattle or poultry sheds are treated with pesticides to destroy pests living in such environments. When such livestock are used for consumption, the toxic residues accumulated in their bodies may find their way into the body of man.

vi) **Affecting other terrestrial animals**. While applying pesticides, some of the toxic materials may fall on grasses and other vegetation. The chemicals drifting and falling into water sheds, as well as the toxic residues found in the vegetation and water sources may affect wild animals and others, which feed on the vegetation and drink from the contaminated water sources. Rodenticides used to kill rats and other animal pests may kill foxes, rabbits, boars etc. The toxic materials reaching the water sources may kill snakes, frogs, toads, crabs etc. as well as animals that drink the water.

vii) **Affecting water - inhabiting organisms**. When pesticide residues reach and contaminate water sources such as, ocean, lakes, tanks, ponds, rivers, canals, water sheds etc., fishes, frogs, tortoise etc., which live in water are affected. When man consumes such organisms, the toxic residues found inside their bodies may enter into the body of man and may affect him.

viii) **Affecting soil - inhabiting organisms and vegetation**. When pesticides are sprayed or dusted, certain quantity may fall on the soil. Further, pesticides are also applied directly to soil to destroy soil - inhabiting pests. The pesticides thus reaching the soil may affect and kill beneficial organisms such as, earthworms, nitrogen fixing bacteria, nitrifying bacteria etc., thereby the quality of the soil may also be adversely affected. Such pesticides may also find their way into the underground parts of plants viz., tubers, rhizomes, bulbs etc. and the toxic residues may be found in such parts even after harvest and when consumed may enter into the body of man.

ix) **Affecting beneficial insects**. Most of the pesticides, besides destroying the insect pests, may kill beneficial insects such as, parasitoids and predators, thereby the balance of life may be upset leading to flare up in some of the serious pests, which may cause extensive damage to crops. Some of the pesticides are toxic to insect pollinators such as, honeybees, wasps, beetles etc. and kill them.

x) **Affecting insect pests**. When some of the insecticides are used repeatedly within a short span of time, some of the insect pests may develop resistance against such insecticides. The insects, which have developed resistance to a particular insecticide cannot be killed even with very high doses of that chemical. Further, subsequent generations of such pests may also possess resistance to that insecticide and as such they may multiply enormously and cause extensive damage to crops. Many instances of such insect resistance have been reported. Rice stem borer, which is a very serious pest of rice, has developed resistance against diazinon. Some insects have been reported to have developed resistance against aldrin. Mosquitoes exhibit resistance to many insecticides.

Some insecticides, though capable of controlling the pests effectively in the initial stages, when used repeatedly may induce resurgence in some of the sucking insect pests, which may multiply in large numbers and cause severe damage to crops. This is attributed mainly to the destruction of natural enemies of the pest, which usually keep the pests under check. When insecticides such as, synthetic pyrethroids , quinalphos, fenthion etc. are used repeatedly to control borer pests of cotton, they induce resurgence in white flies and other sucking insect pests.

### 2. Affecting the Agro - ecosystem

The soil, water and air surrounding the earth are the chief factors of the agro - ecosystem or environment. When the toxic residues enter into the environment, they may affect man and all living beings in one way or another. The toxic residues of chlorinated hydrocarbons, Arsenic compounds, Mercuric compounds and some nematicides remain in the soil and spoil the quality of the soil. Similarly, the toxic residues in the air and water cause air and water pollution respectively and spoil their nature.

### Pesticide residues

The pesticides used for the control of insect and non - insect pests are not disintegrated to form simpler and non - toxic chemical compounds immediately, but may remain in the applied parts and in the environment for several days. Such qualities in pesticides depend upon their physical and chemical nature. The prolonged existence of toxic materials depends upon their ability to undergo various chemical reactions in combination with other elements such as, volatilization, oxidation, reduction, hydrolysis, bacterial degradation etc. The pesticides, which remain as such without undergoing any kind of degradation, are known as **'toxic residues'**. The pesticides, which remain inside the plant tissues without degradation afford protection to the plants against pest infestation. However, the toxic residues found inside the harvested grains, agricultural products, fodder etc. may adversely affect man and other livestock and cause dangerous aftereffects.

In general, the toxicity of organic pesticides last for several days. But the toxic residues of products obtained from plants such as, Pyrethrum, Nicotine, Ryania, Rotenone etc. are degraded within a few days in the presence of heat, light, moisture etc. without leaving behind any toxic residues. The toxic residues of organochlorine compounds persist for several days. The residual toxicity of systemic insecticides after being absorbed and translocated throughout the plant tissues persist in the tissues of plants until they are degraded by hydrolysis or by undergoing chemical reaction with some of the enzymes produced by the plant tissues. The toxic residues of pesticides found outside the plant surfaces are degraded by natural agents such as, heat, light, rain - water, wind etc.

Inorganic pesticides such as, Lead arsenate, Cryolite, Mercuric chloride etc. are not degraded easily and may persist for long periods. Toxic residues may persist in the soil, water and inside the body of living beings also for varying periods.

## 1. Pesticide residues in soil

The toxic residues of chlorinated hydrocarbon insecticides such as aldrin, dieldrin, heptachlor, chlordane etc. may persist in the soil for several days. The pesticide residues, which reach the soil may enter the body of soil - inhabiting invertebrates or mixed with water in water sources or degraded by soil microbes and natural agents. The persistence of toxic residues depends on the physical and chemical nature of the soil, type of formulation of the pesticide etc. The toxic residues of granular formulations may persist in the soil much longer than wettable powder and emulsifiable concentrate formulations. In clay soils, the toxicity may persist longer than in sandy soils. The toxicity of organochlorine and organophosphorus pesticides may last longer in acidic soils than in alkaline soils. Similarly, in soils rich in humus, the residual toxicity persists much longer. When the soil temperature and soil moisture is high, the toxicity does not persist long. In soils where there is high population of microorganisms, the toxic residues are degraded very rapidly.

## 2. Pesticide residues in water

The toxic residues of pesticides reaching water sources settle down to the bottom and are adsorbed by the clay particles. They may also enter into the body of water - inhabiting microbes, insects and vertebrates or into the tissues of algae, water plants etc. or get diluted in water. As such, there may not be much chance of the concentration of toxic residues to increase mostly. The toxic residues persisting in the body of insects, frogs, fishes or such other organisms living in water may adversely affect the birds, snakes, as well as man consuming them.

## Pesticide residues in living organisms.

i) **In plants.** The toxic residues of pesticides used for plant protection purposes may remain either on the surface of the plant parts or inside the pant tissues. It is obligatory that the pesticide residues should not be present beyond a certain limit in food products used for consumption by man and livestock.

ii) **In birds**. The toxic residues present in grains, grain products, vegetation, fishes etc. may enter into the body of several bird species, which feed on such things. When such toxic residues accumulate beyond a certain limit in their bodies, they may die.

iii) **In food products**. When highly toxic and persistent pesticides are applied just prior to harvest or while in storage, the toxic residues may be present in such products and when these are consumed by man or livestock, they may enter into their bodies. Some of the toxic residues of organochlorine compounds are retained in the fat tissues of milk animals and come out through their milk.

iv) **In the body of man**. The toxic residue of DDT was found in the body tissues of man up to a concentration of 5.0 - 10.0 ppm. This chemical has since been banned because of its high residual toxicity. The toxic residues of some of the organochlorine compounds such as, aldrin, dieldrin, heptachlor etc. get accumulated in the human body. These chemicals have also been banned or their use has been very much restricted. The toxic residues of some of the pesticides are found to be passed on from pregnant woman into the body of the child still in the womb of the mother. These residues have been found in the milk of mothers also.

v) **Pesticide residues in food**. Any specified toxic residual substances present in food, agricultural commodities, animal or poultry feeds resulting from the use of pesticides is termed **'pesticide residues'**. Most of the foods we consume although treated with agricultural chemicals are almost free from residues. However, the possibility of harmful residues in food poses a serious concern to the public. When a pesticide is applied to a crop or to soil, it is broken down to simpler, non - toxic substances within a stipulated period of time by the action of various physical, mechanical, biochemical and biological agents such as, light, air, microorganisms and metabolism of plants. In most of the cases, there is a long time gap between the time of application of the insecticide and harvest of the crop and in all possibilities, there is no chance of the toxic substances remaining as such without undergoing degradation in the plant produce. In the case of some pesticides, which may give rise to residue problem, the pre - harvest intervals are stipulated by the approving authorities and are clearly indicated on the instructions.

Before being consumed most of the food and food materials are stored or processed, which also reduce the residues. Heating, boiling, cooking, frying and such other methods of processing help to reduce the toxic residues in the food to non - toxic levels. However, vegetables, greens, tubers etc., which are eaten raw without cooking may contain more of the toxic residues. Many countries adopt legislative measures to regulate the levels of pesticide residues in food materials, which is guided by the principles of **'Acceptable daily intake'** (ADT) and **'Maximum residue levels'** (MRL).

Acceptable daily intake denotes the level of a toxic chemical residue to which daily exposure over a lifetime is not enough to cause appreciable risk. It is expressed in mg. of the chemical per kg. of body weight, at which no adverse effect is observed. Maximum residue levels are estimated from globally generated pesticide residue data.

## Factors affecting effectiveness of insecticides

The effectiveness of insecticides is influenced either directly or indirectly by several factors:

i) **Mode of entry**. Mode of entry of the insecticide into the body of the pest is an important factor in the control of pests. The insecticide may enter into the body through the skin, through the mouth or through the respiratory tracts. Adding an oil to a spray fluid makes it lipophilic and lowers it's surface tension and thus it is able to penetrate through the cuticle more easily and enhance its toxicity. Stomach poisons are absorbed in the midgut easily and interfere with the enzyme activity. In the case of respiratory poisons, the spray enters through or penetrate the tracheal system easily because of the lowered surface tension.

ii) **Developmental stages of the insect pest**. The effectiveness of an insecticide depends to a large extent on the developmental stages of the pest. The early developmental stages of insect larvae or nymphs are often more susceptible to insecticides than the older stages.

iii) **Environmental factors**. Environmental factors such as, temperature, humidity, rains, sunlight, air currents etc. greatly influence the effectiveness of insecticides. Quick - acting insecticides are generally more effective at high temperatures, but get detoxified inside the body of the insect quite easily. On the other hand, slow - acting poisons are more effective at relatively lower temperatures. In the case of concentrate spraying, the spray fluid that comes out as fine mist may dry off quickly under conditions of low relative humidities and high temperatures. Rains may wash off water - soluble sprays. Sunlight may be responsible for the break down of the toxic residues of certain insecticides. Air currents may affect spraying or dusting operations resulting in drifting and uneven coverage. This is more so when spraying or dusting operations are taken up by aircrafts.

iv) **Nature and condition of the plants**. In the case of plants having leaves with waxy coating, the spray fluid may run off without sticking on to the leaf surface. This can be corrected by adding wetting and spreading agents to the spray fluid. In most of the commercial insecticidal formulations, wetting and spreading agents are added to overcome this problem. When there is thick or dense foliage on the plants, the coverage of the insecticide may not be uniform.

## Pesticide poisoning

Use of pesticides has become almost an integral part of agricultural production in recent years. Use of pesticides indiscriminately without any reservation, use of lesser dose of pesticides than the recommended dose of pesticides, not using appropriate pesticide to control the target pest, using the same pesticide repeatedly over a short period of time, using a mixture of several pesticides at the same time etc. have resulted in the development of resistance in the pests, resurgence etc., and above all destruction of natural enemies of the pests, which are comparatively very few in number in nature and are more vulnerable to pesticide poisoning than the pests, thereby upsetting the balance of nature. Though natural farming or organic farming is much better, under the present circumstances it may not be feasible. The change cannot come all of a sudden. It has to come only at a slow pace, which may take more time and till then use of pesticides is inevitable.

Pesticide poisoning through the harvested produce can be avoided to a large extent by stopping pesticide application at the time of maturity of the crop. In the case of moderately poisonous or less poisonous pesticides, harvest of the crop should be done about a week after treatment. In the case of highly poisonous pesticides, harvest of the crop should be taken only about two weeks after treatment, so that the residual toxic substances may disintegrate by the time the crop is harvested.

While using pesticides, poisoning may happen accidentally or deliberately and to avoid such happenings certain precautions have to be taken:

## First aid precautions

Read the instructions furnished in the label of the container carefully before undertaking pesticide application.

i) **Swallowed poisons**. In the case of swallowed poisons, the poison has to be removed from the patient's stomach immediately by inducing vomiting. This can be done by giving common salt - 15 g. in a glass of warm water as an emetic. This should be repeated till the vomit is clear. If the patient is already vomiting , large amount of warm water is given and no emetic should be given. Further treatment should be given according to specific instructions by a qualified physician.

ii) **Inhaled poisons**. The patient should be carried to a place with good aeration and all tight - fitting clothing should be loosened. The patient should be wrapped up in a blanket to avoid chilling. Artificial respiration should be given, if necessary. The patient should be kept calm and quiet. Alcohol in any form should not be given. The patient should be immediately referred to a qualified physician.

iii) **Skin contamination**. In case of accidental spilling of pesticide spray fluid on the clothing or skin, the clothing should be removed and the skin drenched and thoroughly washed with running water immediately.

iv) **Eye contamination**. In case of accidental fall of pesticide spray droplets in the eyes, the eyes are kept open and gently flushed with a stream of running water repeatedly. No chemicals should be applied. The patient should be referred to a eye specialist immediately.

v) **Prevention of collapse**. In case of serious poisoning and collapse, the patient should be covered with a light blanket. The patient should not be exhausted by too vigorous treatments. The foot of the bed should be kept raised slightly. Strong tea or coffee may be given. With the help of a physician, hypodermic injection of stimulants such as, caffeine and Epinephrine may be given. Dextrose 5% solution may be administered intravenously. If necessary. blood or plasma transfusion may be taken up.

In case of any kind of pesticide poisoning it is very important that the patient should be immediately rushed to a qualified physician.

## Antidotes

i) **Removal of poison from the stomach**. This is done either by inducing vomiting or by gastric lavage.

ii) **'Universal antidote'**. Universal antidote is commonly used in case of many kinds of pesticide poisoning. It contains a mixture of Activated charcoal - 7.0 g, Magnesium oxide - 3.5 g and Tannic acid - 3.5 g in half a glass of warm water. It is effective in cases of poisoning by acids, liquid glycocides and heavy metals. It absorbs or neutralizes the poison. Except in cases of poisoning by corrosive substance, it should be followed by gastric lavage.

iii) **Demulcents**. These are substances, which have soothing effect. Demulcents are administered after the stomach contents are emptied as completely as possible. Raw egg white mixed with water, gelatin (9 - 18 g) dissolved in about 500 ml of warm water, butter, cream, milk, mashed potato, flour mixed with water etc. are some of the demulcents.

## Specific antidotes for some pesticides

i) **DDT and other organochlorines**. First of all 'Universal antidote' should be given, followed by gastric lavage. After the stomach has been emptied as completely as possible, Magnesium sulphate (Epsom salt) - 28 g. in a glass of water is given, followed by hot tea or coffee. Then 10% Calcium gluconate - 10 ml. should be injected intravenously. If necessary, Phenobarbital - 0.1 g. should be administered intravenously. To prevent damage, the patient should be fed with rich carbohydrate and Calcium diet.

ii) **Aldrin and Dieldrin**. The patient should be given an emetic and allowed to vomit. This may be repeated if necessary till the vomit is clear. The patient should be allowed to lie down and kept calm and quiet. Then with the help of a physician, Phenobarbital or Barbiturates should be administered intravenously.

iii) **Organophosphorus insecticide**. Organophosphorus poisoning results in blurred vision, abdominal cramps and tightness in the chest. With the help of a physician, Atropine - 2 tablets (each 10 mg.) should be given. In case of respiratory failure, artificial respiration should be given. Morphine should not be given.

iv) **Zinc phosphide**. First of all the patient should be made to drink one teaspoon full of mustard powder mixed in a glass of warm water. This will induce the patient to vomit. After the vomiting has stopped, the patient should be given a solution of Potassium permanganate - 5.0 g. dissolved in a glass of water. After 10 minutes, the patient should be given a solution of half a teaspoon full of Copper sulphate dissolved in a glass of water. Administering a solution of one tablespoon full of Magnesium sulphate dissolved in a glass of water, 15 minutes later should follow this. The patient should be taken to a physician immediately.

v) **Bromodialone**. The patient should be given Vitamin K orally or intravenously with the help of a physician. The treatment should be repeated if necessary.

## Toxicology

**'Toxicology'** is the science of poisons. Toxicology may be defined as 'the branch of medical science that deals with the nature, properties, effects and detection of poison'. Insect toxicology is 'the study of economic poisons, their effects, mechanism of action and metabolism of toxicants in insects'. **'Toxicity'** is the ability of a chemical to bring about changes in the biological system , which are harmful to the target organism. Toxicity may be of two types *viz*., **'acute toxicity'** and **'chronic toxicity'**. Acute toxicity is the acute stage of poisoning due to the application of a single dose of the toxicant. Chronic toxicity is the condition of toxicity, which lasts for the entire life of the target organism as a result of accumulation of small, repeated doses. The response of an organism to a given toxicant is assessed in different ways.

**Median lethal dose ($LD_{50}$).** The dose of the poison required to bring about 50% mortality is known as the median lethal dose. It indicates the general criterion for the acute toxicity of a solid or liquid compound. So, the median lethal dose is the amount of the toxicant required to kill 50% of the test population of insects and test animals and is expressed in terms of milligrams of the substance of the toxicant per kilogram body weight (mg. / kg.) of the test animal, when treated orally. In the case of insects, the $LD_{50}$ value is expressed in terms of micrograms of the toxicant per gram body weight of the insects.

**Acute dermal lethal dose (Dermal $LD_{50}$).** It is the amount of toxicant required to be placed on the skin to cause death of 50% of the test population.

The toxicity of a toxicant shows wide variations to different species of insects and hence separate tests are required to assess the $LD_{50}$ values of each and every species. The higher the value of $LD_{50}$ lesser is the toxic nature of the toxicant. The median lethal dose is assessed by applying different doses of the test toxicant on a pest population on an experimental basis. The percentage of mortality of insects at different doses of the toxicant (X axis) is plotted against the logarithm of the corresponding dose (Y axis). A sigmoid curve is thus obtained. In this dose response graph, it is rather impossible to ascertain the dose giving 100% as well as 0% kill. This is because that at both these situations, the graph becomes more or less parallel to the 'X' axis. This is so because very low or very high doses will not have any appreciable effect on the mortality. On the other hand, the graph is steeper in the region of 50% response. At this stage even the slightest change in the dose will produce substantial change in the mortality. Thus this most sensitive dose, which results in 50% mortality, is referred to as $LD_{50}$ and this is the reliable index of toxicity. It is for this reason that $LD_{50}$ is adopted as the standard for comparing the relative toxicity of different toxicants.

The classification of pesticides on acute toxicity levels is presented in **Appendix - 13** The median lethal dose, both oral and dermal pertaining to some of the important insecticides is furnished in **Appendix - 14.**

**Median lethal concentration ($LC_{50}$).** It is expressed in terms of concentration of the toxicant , which gives 50% mortality of the population of test animal. It is usually determined by a 'Potter's cover test' and ' probit analysis' **$ED_{50}$ and $EC_{50}$.** $ED_{50}$ denotes the dose of chemicals such as, chemosterilants, which brings sterility in 50 % population in the test organisms, while $EC_{50}$ is the concentration of the chemical that results in sterility of 50 % of the population of the test organisms.

## Bioassay of pesticides

**'Bioassay'** is a combination of two words viz., **'Bios'** and **'Assay'**. Bios means life and Assay means determination. Bioassay is the measurement of the potency of any stimulus that induces reactions, either physical, chemical, biological, physiological or psychological in living organisms. In other words, bioassay is the study of response of the individual organism exposed to the toxicant.

Bioassay of pesticides is carried out both under field and laboratory conditions. In the case of field evaluation, the particular chemical is evaluated for its efficacy in controlling the pests under conditions of wide variations, which include weather conditions, crop variety, stage of the crop, degree of infestation, growth stage of the pest etc. In order to get a reliable evaluation, several field trials have to be conducted with different varieties of the crop, in different localities, at different seasons and at different degrees of pest infestation. This involves elaborate recording of observations at different intervals. As such, this process is more time consuming, very cumbersome and quite costly. In this system of bioassay, pest counts are taken before application of the pesticide and subsequently 24, 36 and 72 hours after the treatment and the percentage of mortality of the pest assessed. The mortality of the pests after a span of 2 or 3 days after pesticide application may be considered as a result of residual toxicity.

In the case of laboratory evaluation, the inherent toxicity of the actual amount of toxicant on the test organism is directly evaluated. Such laboratory bioassay is simple, reliable and does not require any sophisticated equipment. Laboratory bioassay is useful for comparing the efficacy of various pesticides, as well as in residue analysis. The time required for such bioassay is also short. A few sensitive living organisms such as, *Drosophila melanogaster, Musca domestica, Bracon brevicornis, Tribolium castaneum* etc., which can be easily multiplied under laboratory conditions are being used for bioassay.

**Charles Potter** in the 1930s introduced the instrument known as the **'Potter's spray tower'** or **'Potter's tower'** for the bioassay of pesticides on test organisms and is still being used in research laboratories all over the world. The instrument, which is used for precision spraying of pesticides, provides a uniform mist over a given area of 9.0 cm. diameter. The target insects are placed on the surface to be sprayed and then sprayed with the pesticide or a residual deposit of the pesticide is provided on the surface and then the target insects are placed on the surface. The target insects are thus exposed for 30 minutes and then shifted to recovery chambers. Mortality counts are recorded 24 hours after treatment. The satisfactory level of mortality is 90 % or more. The residual toxicity is evaluated at different intervals.

Evaluation of **'Space spray'** against flying insects is carried out using another instrument known as **'Peet Gredy Chamber'**.

## Compatibility of plant protection chemicals

When a crop is infested by more than one different types of pests or by a pest and a disease causing organism, it may be necessary to apply different types of insecticides with different modes of action or an insecticide and a fungicide to control them. However, instead of applying the different chemicals separately, if they can be mixed together and applied in a single operation, labour, time and expenditure can be saved to a large extent. When two or more insecticides or an insecticide and a fungicide are mixed together, they may lose their efficacy to control either the pest or the pest and disease. In such cases the chemicals are said to be incompatible. Incompatibility between such chemicals may be either **'Chemical incompatibility'**, **'Phytotoxic incompatibility'** or **'Physical incompatibility',** In the case of 'Chemical incompatibility', when the chemicals are mixed together, chemical reaction may take place and different compounds may be formed, which may not have any effect on the target organisms. Such reactions take place when synthetic organic compounds are mixed with alkaline compounds. In the case of 'Phytotoxic incompatibility', though the individual compounds by themselves are not injurious to plants and do not undergo any chemical reaction on mixing, the mixture may cause injury to plants and the plants may show toxic symptoms such as, blighting of the foliage, leaf fall, shedding of flowers, fruits etc. In the case of 'Physical incompatibility, the compounds may change their physical form and may become unstable.

On the other hand, when two or more insecticides or an insecticide and a fungicide are mixed together, if no chemical reaction takes place and if the efficacy of the individual chemical to control the pests or the pest and the disease is exercised simultaneously without causing any injurious effects on the plants, then the chemicals are said to be compatible. So, it is of great advantage to know the compatibility of plant protection chemicals so that whenever necessary, they can be mixed together and applied simultaneously to control either more than one pest or a pest and a disease.

Spraying a mixture of malathion and carbaryl on cotton plants causes injury to the plants. Bordeaux mixture is not compatible with a number of insecticides such as, endosulfan, methyl demeton, malathion etc. If fungicides such as, captan and difolatan are mixed with insecticides such as, malathion, methyl demeton etc., the mixture should be used immediately after mixing. If they are kept for a longer period of time after mixing, chemical reaction may set in and their efficacy will be lost. It is advisable not to mix antibiotics with insecticides.

# 4

# Plant Protection Appliances

For effective and economic control of pests, besides selection of appropriate plant protection chemicals, application of the chemicals at the most vulnerable growth stage of the pests and use of chemicals at the correct dosage, selection of proper plant protection appliances to carry out the operation is also of vital importance, so as to ensure uniform application of the chemicals.

Plant protection chemicals are mostly used as sprays and dusts using sprayers and dusters respectively. However, in most of the cases spraying is preferred to dusting. Different types of sprayers and dusters are being used for these operations and in both, there are manually operated as well as power operated equipments and in sprayers there are both hydraulic and pneumatic types of sprayers.

The nature of the pesticide formulations to be used and the purpose for which they are used also determine the type of equipment to be used. Dusters are used for dusting insecticidal dust formulations. Sprayers are used for spraying insecticidal liquid formulations. Granular applicators are used to apply granular formulations. Fumigators are used to apply fumigants to fumigate godowns, warehouses, rat holes etc. Soil injectors are used to inject pesticides at a particular depth of soil under pressure. Flame throwers are used to throw flames at specified targets.

## Sprayers

The main function of a sprayer is to atomize or break the pesticidal solution into fine droplets and to ensure uniform distribution of the pesticide in a specific quantity of spray fluid on the surface of the parts to be sprayed such as, plants or soil. Depending upon the type of sprayers, the spray fluid is broken into droplets of size varying from 30 - 400 microns.

Based on the quantity of spray fluid used on a specific area, the spraying is termed **'high volume spraying'**, **'semi low volume spraying'**, **'low volume spraying'** and **'ultra low volume spraying'** and to achieve this, different types of sprayers viz., high volume sprayers, semi low volume sprayers, low volume sprayers and ultra low volume sprayers respectively have to be used.

For **'high volume spraying'**, a quantity of 200 - 300 litres of spray fluid is required to cover one acre of small crops such as, rice, finger millet, groundnut etc. The spray fluid is broken into droplets of size 250 - 400 microns and then sprayed . When several such droplets fall on the sprayed surface, the droplets may aggregate to form bigger droplets and tend to roll down to the ground, thus leading to wastage of a certain quantity of the chemical.

In **'semi low volume spraying'**, which is done with the aid of ordinary power sprayers, the concentration of the spray fluid is 2 - 4 times more than the high volume spraying and one acre can be covered with 50 - 100 litres of spray fluid. The spray fluid is mixed with air and forced out at high speed, as a result the spray fluid is broken into very small droplets of size 150 - 200 microns and sprayed.

In **'low volume spraying'**, which is also called **'concentrate spraying',** the concentration of the spray fluid is 8 - 25 times more than the high volume spraying and consequently the water to be added to the pesticide formulation is much lesser. One acre can be covered with 5 - 50 litres of spray fluid. The spray fluid is sprayed under high pressure and it is broken into minute droplets of size 70 - 150 microns. Here the chances of the droplets aggregating together to form bigger droplets and rolling down the sprayed surface is much reduced or almost eliminated, at the same time more area can be covered in lesser time.

In **'ultra low volume spraying'**, no water is added to the pesticide formulation and it is sprayed as such. To cover one acre, 0.2 - 2.4 litres of the formulation is sufficient. The formulation is mixed with air at high pressure and then sprayed. The chemical formulation is broken up into very minute droplets of size 30 - 50 microns and comes out of the sprayer as a fine mist and settles on the crop surface.

## Advantage of spraying

i) While spraying the chances of the spray being blown off by wind is much less and so wastage of the pesticide, as well as pollution of the environment is reduced to a considerable extent, whereas while dusting, the chances of the pesticide being blown off by wind is much more.

ii) While spraying, the pesticide is distributed uniformly and is dispersed on the parts sprayed and affords better protection against pests, while in the case of dusting the dust is not distributed quite uniformly.

iii) While spraying, the chances of the persons engaged in spraying operation being affected by the toxic effect of the pesticide is much less, while in dusting the chances of the poisonous chemicals affecting persons engaged in dusting operation is comparatively more. The dust particles floating in the air may enter the body through the mouth, through the nose or through the bare skin.

iv) In the spray fluid, the toxic ingredient is held in suspension or dissolved in water, as such the pesticide is distributed uniformly along with the water, whereas in the dust formulations the chances of the toxic ingredient getting separated from the inert fillers is more because of differences in their densities and as such the distribution of the active material may not be uniform and may result in wastage of the active ingredient.

v) Spraying is considered to be about **'4 times'** more efficient than dusting. Though spraying is more widely prevalent than dusting, under some circumstances dusting is more preferable.

**Advantage of dusting**

i) Dusting is more preferable to cover large crop areas, especially in rain fed crops as the cost of dusting is relatively much less.

ii) When water availability is scarce, dusting is preferable.

iii) Dusting is preferable to control pests affecting the earheads of crops. Dusts can be applied even before a short period prior to harvest, as the toxic residues of dusts may not be present in the harvested produce.

iv) Large areas can be covered in less time.

v) For dusting operations less number of persons are sufficient.

vi) Dusters are mostly simple in mechanism and are light in weight and hence dusting is easier.

vii) In hilly and rough terrains dusting is preferable.

viii) The dust formulations may not corrode and cause damage to the equipments.

**Advantages of low volume spraying**

i) In this method much less water is necessary to prepare the spray fluid and as such more area can be covered in less time and the expenditure towards spraying operations is reduced to a considerable extent.

ii) Within a short period of time more area can be covered and because of this, timely spraying may be taken up to cover extensive areas to obtain maximum control of the pests.

iii) Because less quantity of water is required to prepare the spray fluid, the weight of the spraying equipment is comparatively less, which makes spraying operations less tedious.

However, while spraying by this method the chances of the spray fluid being carried away and wasted is more and so when the wind velocity is more than 8.0 KM / hour, spraying by this method should be avoided.

**Parts of a typical sprayer**. A typical sprayer used \for spraying pesticides consists of the following parts :

i) **Fluid tank**. The spray fluid tank is provided inside the sprayer itself in sprayers such as, pneumatic hand sprayers, knapsack sprayers etc., whereas in bucket sprayers, rocker sprayers, foot sprayers etc., the tank for the spray fluid is not provided inside the sprayers. The spray fluid is kept in separate vessels outside into which the suction hose is put. The tank that is provided inside the sprayer itself should be made of non - corrosive metals such as, brass, bronze, or stainless steel, which can withstand the chemical action of the spray fluids or of high density polyethylene material. The tank capacity is one litre or less in small hand sprayers, 10 - 25 litres in backpack sprayers such as, hand operated knapsack sprayers and semi low volume sprayers or power mist blowers and up to 2,700 litres in big trolley mounted or tractor mounted power operated sprayers. The tank is provided with a mouth with screw cap to fill up the spray fluid, a metal strainer to filter the spray fluid while filling up the tank, a tap at the side of the bottom to remove the unused spray fluid and an agitator to constantly agitate the spray fluid at the time of spraying.

ii) **Agitator**. Emulsifiable concentrates and wettable powders, when mixed with water to form a spray fluid, do not go into a solution, but are dispersed in water as finely suspended particles. When the spray fluid is not agitated, these suspended particles may tend to settle down and hence the active ingredient may not be distributed quite uniformly. To prevent settling down of these suspended particles, the spray fluid is constantly agitated while spraying and for this purpose, a paddle - like agitator is provided inside the fluid tank in such a way that when the sprayer is operated, the agitator also rotates inside the tank and agitates the spray fluid constantly. In sprayers, which do not have an inbuilt agitator inside the fluid tank such as, bucket sprayers, rocker sprayers, foot sprayers etc. the spray fluid is agitated frequently manually with a rod or a stick.

iii) **Filters.** Dust, refuse, caked materials etc. found in the spray fluid may clog the nozzle and obstruct spraying operations. To prevent this, a strainer is provided in the mouth of the tank, which filters off all such materials while filling up the tank with the spray fluid. Similarly, finefilters are also provided at the points where the spray fluid passes from the fluid tank into the pump and from the pump into the delivery tube.

iv) **Pump**. The pump helps to force the spray fluid from the fluid tank through the delivery tube under pressure as small droplets on to the area to be sprayed. The efficiency of a sprayer depends to a large extent on the action of the pump.

In pneumatic sprayers, which work on the principle of air pressure, an air pump or pneumatic pump is used. In such sprayers, air is pumped into the tank containing the spray fluid, as a result pressure is applied on top of the spray fluid, which forces the spray fluid to come out of the tank through the delivery tube under high pressure. In such type of sprayers, no pressure is exerted on the spray fluid directly.

On the other hand, in hydraulic sprayers, the pressure from the pump is exerted directly on the spray fluid, as a result the spray fluid is forced out from the pump through the delivery tube as small droplets and then sprayed.

v) **Power source**. For the sprayers operated by hands or legs, only man power is used. For power sprayers, mechanical power from petrol or kerosene powered machines are used.

vi) **Pressure gauge**. In pneumatic sprayers, pressure gauges are provided to measure the air pressure inside the fuel tank or inside the air chamber. Manually operated sprayers such as, Marut sprayers, rocker sprayers, foot sprayers etc., as well as power operated pneumatic sprayers work efficiently at pressures ranging from 15 - 20 pounds.

vii) **Valves**. To regulate the air, or spray fluid entering the pump and to regulate the flow of spray fluid passing on to the delivery tube from the tank, ball - type valves with valve seats are provided. A cut - off valve is also provided in the lance to cut off the spray fluid from being sprayed.

viii) **Hose**. In sprayers where the spray fluid tank is not provided within the sprayer unit such as, the bucket sprayers, rocker sprayers, foot sprayers etc., a suction hose is provided to suck in spray fluid from the vessel into the pump assembly. In all sprayers, a delivery hose is provided to take the spray fluid from the pump to the spray lance to be sprayed. The hoses are made of pliable nylon or rubberized tubes and are made to withstand high pressure, to last longer and to resist the corrosive action of materials such as, petroleum oils, grease and other chemical compounds.

ix) **Spray lance**. It is a long, thin, jointless tube, either straight or smoothly bent slightly at the terminal end and mostly about 90 cm. long in manually operated sprayers and made of non - corrosive metals such as brass. It is attached to the delivery tube with a screw clip and at its terminal end the nozzle is screwed. A handle and a cut - off valve is provided near the place where it is attached to the delivery tube.

x) **Nozzle**. The nozzle is another important part of a sprayer and is responsible to break the spray fluid into small droplets to be sprayed. Several types of nozzles are used to meet different requirements. Depending upon the type of nozzles, the size of the spray droplets and the quantity of spray fluid sprayed vary.

In general, the nozzle consists of a basal bracket, a body, a circular disc or orifice plate, washer, a whorl plate and a screw cap with a central hole. The nozzles are of two types functionally viz., the hollow cone nozzle and the solid cone nozzle.

In the hollow cone nozzle, the spray fluid, which comes out of the pump under pressure, strikes against the whorl plate, swirled and comes out as a hollow boom or hollow cone with its central part hollow, whereas in the solid cone nozzle, the spray fluid comes out through the orifice plate and the whorl plate, as such the hollow of the cone is also filled up with the spray fluid, thus forming a solid cone.

xi) **Air chamber**. In some of the pneumatic, manually operated sprayers such as, rocker sprayers and foot sprayers, a separate air chamber is provided. When the pump is operated, air fills up the air chamber and pressure is developed inside the chamber. The compressed air inside the air chamber forces the spray fluid through the delivery tube. In such sprayers, as long as there is sufficient pressure inside the air chamber, the spray fluid comes out continuously.

## Types of sprayers

## Manually operated hydraulic sprayers

1. **Hand syringe** or **Garden syringe**. It is the oldest and simplest type of sprayer consisting of a long cylinder or barrel made of tin or polythene and a plunger with a wooden handle. At the anterior end of the plunger, a leather valve is attached in - between a metal washer and screw nut. The leather valve fits closely within the wall of the barrel. At the anterior end of the barrel, a nozzle with a small hole is provided. The spray fluid is kept in a separate vessel. It works as an ordinary injection syringe. The nozzle is placed inside the spray fluid and the plunger is pulled up, as a

result a vacuum is created inside the barrel and the spray fluid enters the barrel to fill up the vacuum. Then the sprayer is taken out with the spray fluid inside and the nozzle is pointed towards the area to be sprayed and the plunger is pushed down. The pressure exerted directly on the spray fluid forces it to come out through the nozzle as droplets. Thus every time the spray fluid is drawn into the barrel and then sprayed. The spray droplets are fairly large in size and the spraying is not continuous It is used to spray potted plants and small kitchen gardens **(Fig. 10)**

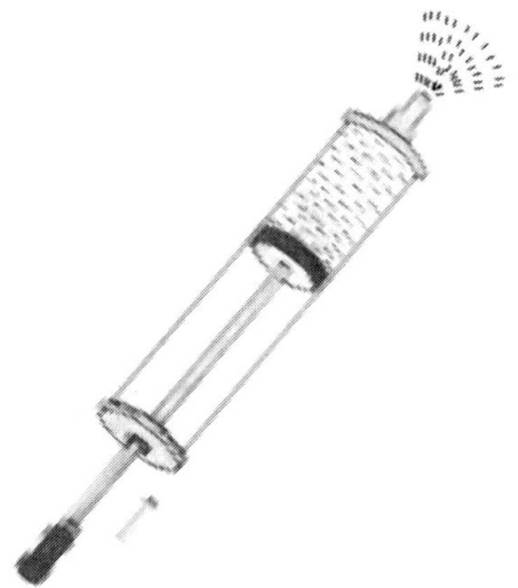

**Fig. 10.** Garden syringe

2. **Bucket sprayer** or **Stirrup pump**. It consists of a long barrel made of brass, a plunger, valve and a foot pedal to hold the sprayer pressed to the ground in an upright position. There is a suction hose with a valve at the side near the bottom of the barrel and a delivery tube at the side near the top end of the barrel, which is closed at the top end. A valve is provided at the delivery tube to which the delivery hose is attached. The spray fluid is kept in a suitable bucket or vessel. The barrel is placed inside the bucket and the sprayer is held in an upright position with the foot pedal. During the upward stroke, the lower valve opens and the spray fluid enters the barrel to fill up the vacuum created. During the downward stroke, the lower valve closes and due to the pressure exerted on the spray fluid inside the barrel, the upper valve opens and the spray fluid is forced through the delivery tube and sprayed through the nozzle. Spraying is done only during the downward stroke and as such the spraying is not continuous. Two workers are required to undertake spraying operation, one to operate the sprayer and the other for spraying. It is suitable to spray small crop areas, kitchen gardens, small bushes etc. **(Fig. 11)**

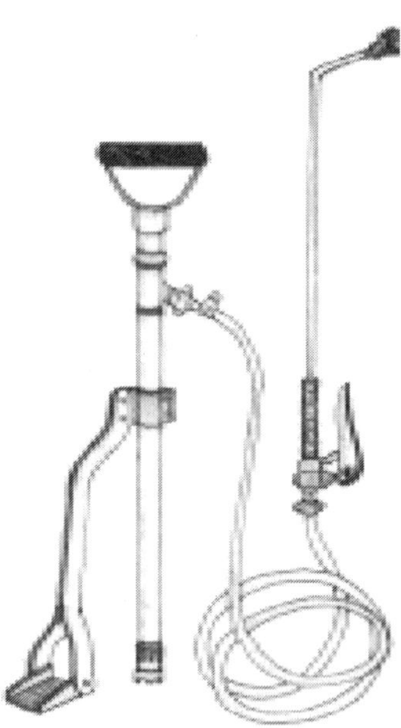

**Fig. 11.** Bucket sprayer

3. **Hydraulic knapsack sprayer**. This type of sprayers can be conveniently carried on the back and for this the tank is shaped like a bean seed with a slight curve. It works on the same principle as the bucket sprayer, but the pump is fitted inside the tank itself. The capacity of the tank may vary from 10 - 20 litres and is made of brass or high - density polyethylene material. With the help of a lever and handle attached to the plunger, it can be operated with one hand. An agitator is also provided inside the tank. When the plunger is moved up and down by operating the lever, the agitator also moves up and down and agitates the spray fluid. There is a delivery tube at the top of the barrel to which a short delivery hose is attached and the spray lance is attached to the terminal end of the delivery hose. A cut - off valve is provided in the lance. One worker is enough to operate the sprayer, by working the plunger with one hand and do the spraying with the other hand simultaneously. Because the plunger is continuously operated up and down, the spraying is done continuously. It is ideal for spraying rice, vegetable crops and other small crops. An area of 1.0 - 2.0 acres can be covered in a day **(Fig.12)**

**Fig. 12.** Hydraulic knapsack sprayer

4. **Rocker sprayer**. This type of sprayers can be operated by standing and moving the long rocking lever forward and backward. It consists of a pump, pressure chamber or air chamber, suction hose, delivery hose, lance, nozzle and valves and the whole sprayer assembly is mounted on a wooden board. There is no tank attached to the sprayer assembly and so the spray fluid is kept in a separate vessel in which the suction hose is put.

   When the pump is operated by the rocking movement of the handle, pressure is developed inside the pressure chamber and the spray fluid, which is sucked into the pump through the suction hose is forced through the delivery tube due to the pressure developed in the pressure chamber and sprayed.

   However, to get a continuous and uniform spray, the pump has to be operated continuously. To operate the sprayer, two workers are required, one for operating the pump and the other for spraying. A long delivery hose may be provided for this sprayer so that larger area may be covered without shifting the sprayer often. Tall crops and fairly high trees can be sprayed with this sprayer by using a straight high tree lance. In the pressure chamber, an air pressure up to 14 - 18 kg./cm$^3$ develops. An area of 3 - 4 acres can be covered in one day **(Fig.13)**

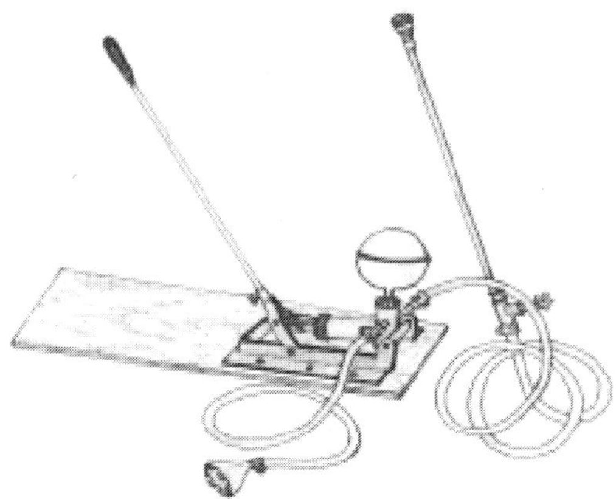

**Fig. 13.** Rocker sprayer

5. **Foot sprayer** or **pedal pump**. This type of sprayer can also be operated from the standing position by foot by operating the foot pedal up and down. This sprayer, which is fitted to a metal stand in an upright position consists of a vertical pump, a plunger, which is attached to a lever and pedal, pressure chamber at the top of the pump barrel, suction tube, delivery tube, spray lance, nozzle, valves etc. The pedal attachment is provided with springs, so that when the pedal is pushed down, it automatically comes up to its original position. There is no tank attached to the sprayer and the spray fluid is kept in a separate vessel into which the suction tube is put. It works on the same principle as that of the rocker sprayer. In the pressure chamber, pressure develops up to 17 - 20 kg. / $cm^3$. Long delivery hose may be attached to the sprayer. Tall crops, high trees etc. can be sprayed with this sprayer by using a high tree straight spray lance. Two workers are required to undertake spraying operations, one for working the pump and the other for spraying. An area of 3.0 - 5.0 acres can be covered in one day **(Fig. 14)**

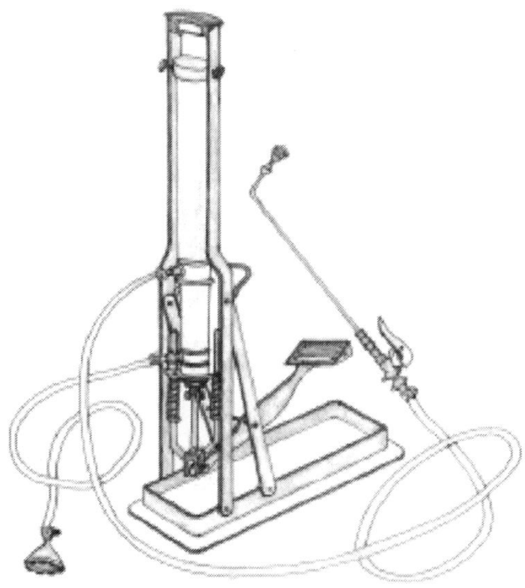

**Fig. 14.** Foot sprayer

## Power operated hydraulic sprayers

In these type of sprayers, which are mostly semi low volume sprayers, the spray fluid under pressure developed by the pump is forced out through the nozzle in small droplets, 150 - 300 microns in size. The sprayer consists of an engine, pump, a tank with agitator, framework with or without wheels, pressure regulator, valves, pressure gauge, suction and delivery hoses and lance with a nozzle. The sprayers are powered by 1.0 - 1.5 HP. air - cooled petrol engines. Different types of spray booms with equidistantly fixed nozzles can be attached to the sprayers. The discharge capacity of these sprayers range from 7.0 - 18.0 litres per minute at pressures up to 18 kg. / $cm^3$. These sprayers are mostly used to spray crops in rough terrains , tea and coffee estates, as well as in rubber plantations, orchards etc., where the spray jet has to reach great heights of 30 - 45 feet or more. There are different types of power operated sprayers

i) **Stretcher type** or **Skid type**. In this type of sprayers, the entire sprayer assembly is fixed on a stretcher - like iron frame with a pair of handles on both the sides of the chassis. The sprayer can be carried by two men like a stretcher from one place to another. They do no have built - in spray tanks. They are mostly provided with long delivery hoses to cover large areas without shifting the sprayer from one place to another often.

ii) **Wheel type**. In this type of sprayers, the entire - sprayer assembly is fitted to a metal chassis, which is mounted on pneumatic or solid rubber wheels. Because the sprayer is mounted on wheels, it can easily be moved from place to place. These sprayers also do not have built - in spray tanks.

iii) **Wheelbarrow type**. These sprayers are too heavy and bulky and so are mounted on wheelbarrows. They are provided with a spray tank attachment with a pump and so the use of separate drum for carrying the spray fluid is eliminated. The tank capacity varies from 50 - 80 litres.

iv) **Tractor mounted sprayers**. In this type of sprayers, the power for operating the sprayer is obtained directly from the tractor engine. The spray fluid tank is also mounted on the tractor. Spraying can be done while the tractor is in motion.

## Manually operated pneumatic sprayers

1. **Pneumatic hand sprayer**. In this type of sprayers, the fluid tank serves as the pressure chamber as well. The spray fluid is forced our through the delivery tube only because of the air pressure built in above the spray fluid and hence some space should be left empty in the fluid tank after filling it with the spray fluid. The tank should not be filled up completely with the spray fluid. A pressure gauge is fitted to the tank to measure the built - in pressure. When the plunger is drawn up, air from outside enters the pump and when the plunger is pushed down , the air inside the pump is forced into the fluid tank and fills the empty space above the spray fluid. When the pump is worked in this way several times, sufficient pressure is built up in the empty space above the spray fluid, as a result the spray fluid is forced out through the delivery tube when the cut - off valve provided in the spray lance is opened. The cut - off valve with the lever arrangement helps to open and close the delivery tube, one end of which opens at the bottom of the fluid tank for the spray fluid to enter the delivery tube and the other end opens at the nozzle for the spray fluid to be sprayed. When the cut - off valve is opened by pressing the lever, the pressure built in inside the fluid tank forces the spray fluid to come out through the delivery tube and sprayed through the nozzle. As long as there is enough pressure inside the fluid tank, spraying can be done continuously. When the pressure becomes less, the pump is worked again to increase the pressure. These sprayers can be used to spray garden plants, flower plants, kitchen gardens and in the house to destroy household insects. The tank capacity ranges from 0.5 - 1.0 litre in such sprayers.

2. **Marut backpack sprayers.** In this type of sprayer also, the fluid tank serves as the pressure chamber. To withstand high pressure, the tank is made of strong brass sheet. In the centre of the tank, a pump is fitted with a plunger and wooden handle in a perpendicular position. A mouth with a screw cap to fill up the tank with the spray fluid and a pressure gauge are provided at the top of the cylindrical tank. A delivery tube is provided at the side near the bottom of the tank. One end of a delivery hose is attached to the delivery tube, while the other end is attached to the spray lance with a nozzle and cut - off valve. It works on the same principle as that of the pneumatic hand sprayer. The tank capacity varies from 3.5 - 17.0 litres and pressure ranging from 2,0 - 4.0 kg. / $cm^3$. is developed inside the tank. The sprayer is placed on the ground and the pump is worked so that sufficient pressure is built in inside the tank. Then the sprayer is carried on the back and spraying is done. One worker is enough to carry out spraying operations and an area of 1.0 - 2.0 acres can be covered in a day. It is used to spray small crop areas and is largely used by the Public Health Department for mosquito eradication programs **(Fig.15)**

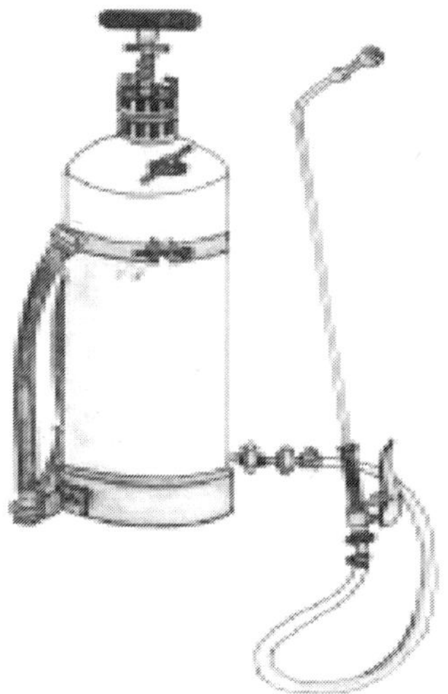

**Fig. 15.** Marut knapsack sprayer

## Power operated pneumatic sprayers

Just as power operated hydraulic sprayers, different types of power operated pneumatic sprayers are also available:

i) **Portable power operated pneumatic sprayer**. In this type of sprayers, compressed air from a compressor driven by a small engine is forced into the spray fluid container having a capacity of about 45 litres. An outlet from the container is attached to lances or booms with nozzles.

The pressure exerted on the spray fluid in the tank forces the spray fluid to come out through the nozzles as minute droplets. Such sprayers are mounted like portable power operated hydraulic sprayers on skids, wheelbarrows etc.

ii) **Tractor mounted pneumatic power sprayers**. Sprayers of this type have much larger tanks of 250 - 2,000 litres capacity and more powerful compressors, which are mounted on tractors and can be used for spraying extensive areas of crops in the plains, orchards etc.

## Manually operated mist blowers

**Hand atomizer** or **Flit pump**. The working of this sprayer is almost similar to pneumatic sprayers. It is a very handy sprayer used for the control of mosquitoes and other household insects . It is also used for spraying potted plants, small kitchen gardens etc. The spraying is not continuous. It is vey useful for spot application in small areas.

The sprayer consists of a cylindrical tube or barrel 25 cm. in length and 3.75 cm. in diameter and made up of tin sheet or high density polyethylene material. The posterior end of the barrel is covered with a screw cap known as the barrel cap, which is provided with a hole in the centre through which passes the plunger. The closed anterior end of the barrel is provided with a small hole in the centre, which is the nozzle . The plunger is made of a metal rod with a wooden handle at the posterior end. A metal washer and a screw nut with a leather valve in - between them, which fits closely with the wall of the barrel serves as a piston. The spray fluid tank is also made of tin sheet or polyethylene material having a capacity of 227 ml. and is provided with a screw cap known as the filler cap. The barrel is attached to the filler cap of the tank on the lower side of the anterior end. A small hole of 3.12 mm. diameter passes through the centre of the filler cap and is at right angle to the barrel. This is the delivery tube, one end of which opens near just half in front of the nozzle, while the other end opens near the inner bottom of the tank. The spray fluid is filled in the tank after removing the filler cap and then screwed back tightly.

When the plunger is drawn back, a good quantity of air enters the barrel through the nozzle. When the plunger is pushed down, pressure is exerted on the air inside the barrel. As the air cannot escape through the piston, it is forced out through the nozzle at great force. When the air goes out with such great force, some of the air inside the delivery tube is drawn out causing a vacuum in the delivery tube, which is filled up with the spray fluid coming up from the tank. This fluid thus entering the delivery tube is also thrown out along with the air in the form of a fine mist **(Fig. 16)**

**Fig. 16.** Flit pump

## Power operated low volume sprayers

**Power mist blower** or **Blower type sprayers**. In such sprayers, the spray fluid comes out of the sprayer as very minute droplets like mist. The size of the droplets vary from 50 - 150 microns and as such the droplets aggregating together to form bigger droplets and rolling down from the sprayed parts and getting wasted is avoided. Water required to prepare the spray fluid in such sprayers is 20 - 80 per cent less compared to the conventional high volume sprayers and as such the concentration of the chemical in the spray fluid is proportionately very high. However, because the spray fluid is micronized into very fine droplets and sprayed uniformly on the crop surface, no spray injury is caused to the sprayed parts. Though the quantity of water to be mixed with the pesticide formulation is much less, the quantity of chemicals to be used should be calculated only on the basis of the water required for spraying with a high volume sprayer.

In some of the low volume sprayers, an additional fan - like atomizer is also attached near the exit point of the delivery hose. The air coming out of the delivery hose at great force makes the fan attachment to rotate very fast and as a result, the spray droplets coming out through the delivery hose are further micronized into still finer droplets. In such type of sprayers, a quantity of 6.8 - 42.5 $m^3$ of air is forced out through the delivery hose every minute at a speed of 200 - 420 km. / hour. The blowers are light in weight, 12 - 20 kg., inclusive of all accessories. This type of sprayers are often used for mass ground spraying for the control of groundnut red hairy caterpillar, rice brown plant hopper, rice green leaf hopper etc., where large areas have to be covered within a short span of time.

In such sprayers, two separate units are mounted on a single metal stand . One is the **'motor'** and the other is the **'sprayer unit'**.

i) **Motor**. The motor of the power mist blower is mostly run on petrol and is powered by a two stroke engine with 1.2 - 3.0 Horse Power and rotates at 5,500 revolutions per minute. When the motor runs at such high speed, the fan attached to the crank also rotates at such high speed, drawing air from outside and forces it out through the delivery hose at very great force. The delivery hose is connected to the fan chamber.

ii) **Sprayer unit**. The sprayer unit consists of a cylindrical fluid tank of 10 - 12 litres capacity placed horizontally. A small petrol tank is also situated by the side of the spray fluid tank. The tanks are made of high density polyethylene material. The fluid tank is provided with a mouth and screw cap for filling the tank with the spray fluid and a delivery tube at the bottom with a tap to regulate the flow of the spray fluid. A thin plastic tube attached to the delivery tube opens inside the posterior end of the delivery hose. The petrol tank is also provided with a mouth and screw cap and a delivery tube with a tap. A thin plastic tube is connected to the petrol tap and the other end of the tube is connected to the carburetor of the engine. The knapsack mist blower commonly used in our country delivers 6.8 - 42.5 $m^3$ of air per minute at a velocity of 200 - 420 km / hour.

When the motor is run, air from the air chamber comes out through the delivery hose with great force. The spray fluid, which comes out through the delivery tube by gravitational force into the delivery hose in small quantities is micronized into small droplets by the air current and comes out as mist, and distributed uniformly on the surface to be sprayed. With the help of an adjustable knob provided at the exit point of the delivery tube. the discharge of the spray fluid from the tank into the fan chamber can be adjusted according to the requirements. An area of 7.5 - 10.0 acres can be covered in one day with such type of sprayers.

When the motor is run, air from the air chamber comes out through the delivery hose with great force. The spray fluid, which comes out through the delivery tube by gravitational force into the delivery hose in small quantities is micronized into small droplets by the air current and comes out as mist, and distributed uniformly on the surface to be sprayed. With the help of an adjustable knob provided at the exit point of the delivery tube. the discharge of the spray fluid from the tank into the fan chamber can be adjusted according to the requirements. An area of 7.5 - 10.0 acres can be covered in one day with such type of sprayers.

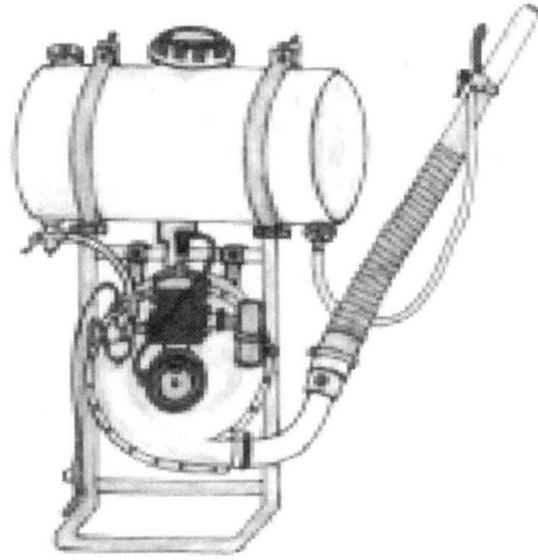

**Fig. 17.** Power mist blower

## Ultra low volume applicators

Ultra low volume applicators or sprayers can be used to spray pesticide formulations as such without mixing with water. However, all available formulations cannot be sprayed like this. Only special formulations, formulated exclusively for such kind of application alone can be used with such sprayers. Such formulations should have low volatility and low phytotoxicity even at high concentrations and only a few chemicals are formulated to be sprayed like this. One acre can be covered with 0.5 - 2.4 litres of the formulation. The spray fluid is forced out through the nozzle as very minute droplets of size 30 - 50 microns like a mist and settles down on the parts to be sprayed uniformly due to gravitational force. Up to 20 acres can be covered in one day with such ultra low volume ground sprayers, as against 7.5 - 8.0 acres with a low volume power operated knapsack sprayer and 1.25 - 2.0 acres with a high volume manually operated sprayer. This kind of spraying is possible with very high pressure pneumatic sprayers provided with special type of spray nozzles and by aircrafts fitted with special type of spray equipments.

## Aircraft application of pesticides or Aerial spraying or dusting

Special types of aircrafts are used to carry out plant protection operations such as, spraying or dusting of pesticides, weedicides, insect repellents, insect attractants etc. Vast areas, especially continuous stretches ranging from 500 - 2,000 acres can be sprayed or dusted in one day. This kind of application can be carried out to control the pests at their most vulnerable stages of development, when it is much easier to control them. Further, within a short period of time vast areas can be covered. The expenditure incurred for aerial spraying is also comparatively less than ground spraying per unit area. Tall crops, fruit trees, high trees, plantations etc. can be sprayed or dusted easily and effectively. Inaccessible areas such as, dense forests, hilly and rugged terrains, marshy places, vast deserts, which are the breeding sites of desert locusts etc. are amenable for aerial spraying or dusting.

However, best results in aircraft applications are obtained under conditions of low wind velocity and high humidity. Low wind velocity prevents drifting and consequent wastage of chemicals, as well as pollution of nearby crops and water sources. Humidity influences rate of evaporation thereby affects uniform distribution of the spray chemical. Dusting is preferably done when the wind velocity is less than 5.0 km, / hour, while spraying can be done satisfactorily even at a wind velocity of 16 km. /hour.

**Fig. 18.** Aerial spraying

Fixed wing aircrafts and helicopters are used to apply plant protection chemicals. Fixed wing aircrafts fly faster than helicopters and can carry more loads. The 'beaver type aircraft', which is commonly used for such operations can cover up to an area of 1,000 acres per 6 flying hours when fitted with low volume spray nozzles and up to 3,000 acres in 6 flying hours when fitted with ultra low volume spray nozzles. They are capable of carrying up to 630 litres of spray fluid in the fluid tank and are powered by 220 - 450 horse power engines. About 30 nozzles are fixed on to a boom. The nozzles used in aircrafts produce hollow cone or flat fan type spray. Application is done at speeds varying from 105 - 180 km. / hour and at heights ranging from 3.0 - 11.0 meters above the ground level. The swathe covered in each sortie is 15 - 60 meters.

Fixed wing aircrafts meant for dusting plant protection chemicals can carry 100 - 1,000 kg. of the dust formulation. The discharge rate of the dust can be controlled by adjusting the gate valve of the hoppers and is calibrated in cu. m. of dust formulation per minute. The discharge rates vary from 0.5 - 26.0 kg. of dust formulation per acre.

Fixed wing aircrafts meant for application of plant protection chemicals should be able to fly at relatively low speed, should have high rate of climb, should posses high load carrying capacity and should have the lowest possible landing speed so that they can be operated in short runways.

On the other hand, the helicopters used for the application of plant protection chemicals can cover an area of 50 - 850 acres in 6 flying hours, when fitted with conventional low volume spray nozzles and up to 2,000 acres when fitted with ultra low volume spray nozzles. They are capable of carrying 180 - 270 litres of spray fluid.

The helicopters meant for dusting are provided with two hoppers of about 85 kg. each on either side of the fuselage. The rate of flow of dust is regulated by means of sliding valves. The swathe covered in each sortie is about 13 meters wide. They can fly at a height of 1.26 meters above the ground level at a speed of about 37.5 km. / hour.

## Dusters

The equipments used for dusting dust formulations directly without mixing them in water are known as **'dusters'**. Both manually operated and power operated dusters are available. In general the dusters have a hopper or dust chamber to carry the dust formulations, an agitator to prevent caking of the dusts and to feed the dust formulation uniformly through the exit hole provided in the dust chamber to the fan chamber, a metering mechanism to regulate the quantity of dust to be dusted, a delivery hose or lance through which the dust is blown out from the fan chamber and a reflector to direct the blast of dust on to the parts to be dusted. A high speed rotary fan in the fan chamber or leather bellows produces a continuous blast of air, which carries the dust along with it with force in an air - borne state.

## Manually operated dusters

1. **Hand rotary duster**. These dusters consists of mainly 4 parts viz., the hopper or container, gearbox assembly, blower unit and the discharge unit. Two types of hand rotary dusters are commonly used for dusting operations. One is round and drum - like in appearance in which all the parts are completely assembled as a single unit. This type of duster is fitted on to the chest of the operator with leather straps. The duster is operated by rotating the handle connected to the gear assembly with the right hand, while dusting is done with the left hand.

2. **Orient hand rotary duster**. This type of duster is rather long and consists of two separate units, the hopper and the blower unit connected by a delivery tube from the hopper to the fan chamber. This is operated by fitting the duster on the left side of the body below the arm pit, with the hopper at the back and the blower assembly in front of the body with leather straps. The gear assembly is operated with the right hand by rotating the handle provided, while dusting is done with the left hand, The working principle of both the types of dusters is quite similar.

   The hopper or container is drum - shaped with a lid and has a capacity of 5.0 - 6.0 kg. A chest plate is provided to attach the duster on to the chest. The crank shaft passes through the middle portion of the hopper and carries two iron paddles on opposite sides of the crank shaft called agitators, which agitate the dust contained inside the hopper regularly so that the passage of dust from the hopper to the blower chamber is not blocked. On the lower portion of the gearbox below the main crankshaft is the fan shaft, which rotates the fan. When the main crank shaft is rotated by rotating the crank, the fan shaft rotates. much faster due to the gear arrangement. An exit hole is provided in the hopper to the fan side, which is called the feeder. The feeder supplies the dust to the blower chamber. A feed regulator is provided on the outer side of the feeder hole to regulate the supply of dust to the fan chamber.

   The gearbox is located on the right side of the hopper and consists of 4 gears. The main gear, which is the largest one, is connected to the main crank shaft and the other 3 gears are serially attached. For rotating the main gear, an iron crank with a wooden handle is provided. The blower chamber or the fan chamber is a small closed chamber and opens at the anterior through a small projecting opening about 2.5 cm. in diameter. A fan with 5 blades is located in the middle of this chamber. A wire netting is provided on the left side of the blower chamber to let in air from outside. The main gear connected to the main shaft engages the serially arranged smaller gears and the smallest gear is attached to the fan shaft. When the main gear is rotated by rotating the main crank shaft, the small gear attached to the fan shaft rotates much faster and the fan attached to this gear shaft also rotates at a much faster speed. The fan sends out the air from the centre of the fan chamber forcibly through the opening in the chamber, as a result a vacuum is created, which is filled up by the dust coming from the hopper into the air chamber and is also blown out along with the air. The exit tube of the fan chamber is connected to a rubber hose with a nozzle at the anterior end, which is covered up with the reflector at the top so that the dust coming out of the blower chamber is directed on to the parts to be dusted **(Fig 19)**

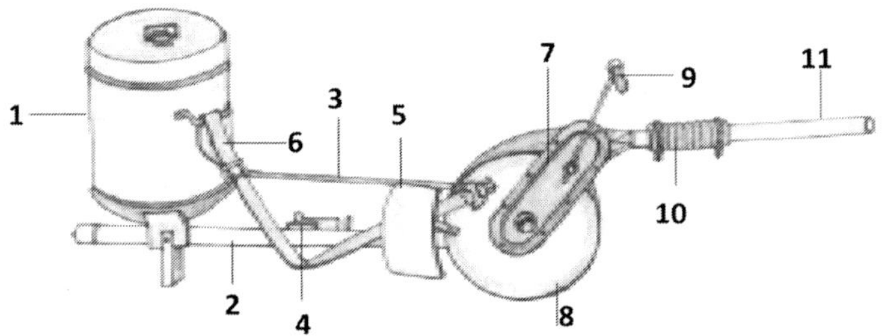


**Fig. 19.** Orient hand rotary duster

1.Hopper 2.Suction tube 3.Agitaor shaft 4.Dust regulator 5.Breast plate 6.Shoulder belt 7.Gear box 8.Fan chamber 9.Crank with handle 10.Delivery hose 11.Delivery tube

Before staring the dusting operation, the hopper is filled up three fourth of its capacity with the dust formulation and the lid is tightly closed. The duster is then tied on to the chest or shoulder with the leather straps. Then the handle is rotated so the gears and the fan move fast along with the agitator. The dust mixed with air is blown out through the lance. To ensure uniform dusting, the operator should move along slowly forward rotating the handle continuously with the right hand and dusting with the left hand.

3. **Bellows duster**. It consists of a small container to hold the dust formulation with a projecting short delivery outlet at the side near the top and a leather bellows on the side of the container. A small hole is provided at the side near the bottom of the container connecting the bellows. When the bellows is worked, blasts of air from the bellows is forced into the container. The blasts of air carrying the dust comes out of the delivery tube.
4. **Power duster**. It consists of a blower for producing air blast, a hopper for holding the dust formulation, a regulator to regulate the supply of dust, flexible hose with nozzle etc., all mounted on movable frames or trolleys along with an engine, which runs the blower and produces air blast. The dust comes out of such dusters at speeds ranging from 300 - 320 km. / hour. The amount of dust to be dusted can be regulated by means of a regulator.

## Aerosols

An aerosol or fog is a system of dispersal of spray droplets of diameter 0.1 - 5.0 microns in compressed gas. Liquefied gas aerosol generators containing insecticides are available in canisters or aerosol bombs for the control of household insects. The aerosol bomb has a thick - walled metal container or canister with a valve and a nozzle. The container contains a liquified gas in which the toxicant dissolved in a solvent is added. The liquified gas produces the pressure and atomization on release into the air. Once the valve is pressed open, the pesticide comes out through the nozzle along with the gas in the form of a fog. After the solvent evaporates, the pesticide remains in the atmosphere in a finely dispersed state. Aerosols are useful for producing small amounts of fog indoors. Propoxur is quite commonly used as aerosol for the control of household pests.

## Other plant protection appliances

1. **Rat fumigation pump** or **Cyanogas pump**. It is used to introduce Calcium cyanide dust into rat burrows, which on coming into contact with atmospheric moisture undergoes chemical reaction and releases the highly poisonous hydrogen cyanide fume. Its action is similar to that of an air pump. It consists of a pump or cylinder made of brass tube measuring 45 cm. in length and 3.5 cm. in diameter. The top of the tube is provided with a screw cap through the centre of which passes the plunger with a wooden handle at the upper end and a leather valve at the other end connected to a washer like a cycle pump. A glass bottle of 0.5 kg. capacity is attached to the lower end of the cylinder. A valve is provided between the pump cylinder and the bottle, which allows compressed air from the pump into the bottle, but prevents it from getting back into the cylinder. A small delivery tube is provided at the side near the base of the pump cylinder and is connected to the glass bottle through which the chemical is forced out. A plastic delivery hose is connected with the delivery tube , which is connected to a 30 cm. long brass tube at the other end. This is known as the lance.

   Before the operation, the glass bottle is filled three - fourth with Calcium cyanide dust and then connected to the pump. The equipment is placed on the ground near the rat burrow and held in position with the foot hold, which is provided below the glass bottle. The lance is introduced into the rat burrow and the plunger is drawn up. When the plunger is pushed down, the chemical along with air is forced into the rat burrow and spreads all over the burrow. Immediately after pumping in the chemical, all exit holes are plugged with clay so the fumes may not escape out. A quantity of 6.0 - 10.0 g. of Calcium cyanide is pumped into each burrow. Other mammalian pests, which live within burrows can also be killed by using this pump **(Fig. 20).**

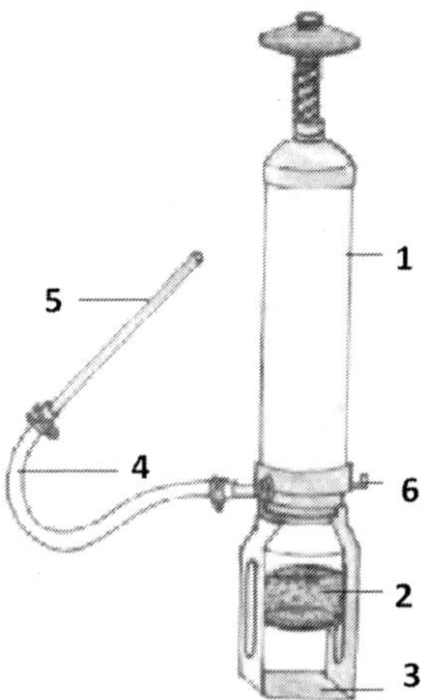


**Fig. 20.** Cyanogas pump

1. Air pump 2. Glass bottle 3. Foothold 4. Delivery hose 5. Delivery tube 6. Agitator

2. **Soil injector** or **Soil gun**. It is an equipment used for soil fumigation with volatile liquids such as, Carbon disulphide, Vapam etc., especially for the control of soil nematodes. It consists of a cylindrical brass tank with an air pump at the centre with a plunger and wooden handle and a long tube with a pointed needle. To penetrate the soil to the required depth, a separate handle is fixed to the tank to push the nozzle of the equipment into the soil and a metering device to regulate the quantity of the fumigant to be injected into the soil. The nozzle tube is 22 cm. in length and a Sliding circular disc is fitted to the nozzle tube with which the depth of application can be adjusted. Several small holes are provided around the pointed end of the of the hollo nozzle through which the fumigant is pumped into the soil. When the equipment is pushed down, the nozzle tube penetrates the soil up to the circular disc. After pushing down the equipment to the required depth and keeping that in that position, the plunger is pushed down,when a metered quantity of the fumigant is injected into the soil. After injecting the fumigant into the soil, the plunger comes back to its original position due to the spring arrangement. Thus each time the plunger is pushed down, a specific quantity of the fumigant is injected into the soil. The quantity of chemical to be injected each time can be adjusted by means of the metering device.

   The tank of the equipment has a capacity of 1,0 - 3.0 litres and the fumigant is injected at a depth of 12 - 22 cm. and to ensure an uniform distribution of the fumigant in the soil, the chemical is injected at different points 30 cm. apart. An area of 1.0 - 1.5 acres can be covered in one day **(Fig. 21)**

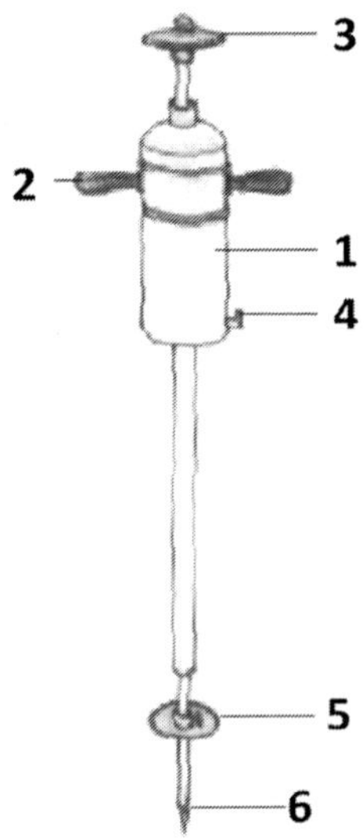


**Fig. 21.** Soil injector

1. Chemical tank 2. Handle for pushing the injector into the soil 3. Plunger 4. Regulator 5. Depth controlling disc 6. Nozzle

3. **Granule applicator**. This equipment is used to apply granular formulations of pesticides uniformly in the fields or in planting furrows or for spot application of granules in particular sites. There are two types of applicators. One is a small hand applicator consisting of a drum - like tank of about 1.0 kg. capacity. A small nozzle is provided at the bottom of the tank through which the granules are applied. Near the base of the tank above the nozzle, there is a provision to slide a circular disc of the diameter of the tank with several holes slightly bigger than the size of the granules, through which the granules are fed into the nozzle by gravitational force and then applied through the nozzle. According to the size of the granules to be applied, the disc can be changed. This equipment is highly useful for spot application and whorl application. The second one is knapsack type and is a bigger one with a hopper of about 10 kg. capacity. It is used to apply granules over larger areas. The equipment can be carried on the back of the operator like a knapsack sprayer.

4. **Bird scarer**. Bird scarers have been deviced to produce loud, exploding sounds at regular intervals to scare away birds damaging crops, grains or fruits. The explosion results from the combustion of Acetylene gas generated by chemical reaction of Calcium carbide with water. In this device, Calcium carbide and water are stored in separate containers in such a way that the water drips from a small container on the Calcium carbide kept in a separate chamber. The pressure developed by the acetylene gas generated causes combustion and production of loud and cracking noise. The rate of flow of water is regulated, which also regulates the time interval of the explosion. The commercially available equipments are designed to produce two explosions per minute or one explosion for every five minutes. The scarer should be hung from the branch of a tree or a tall pole at a height of 20 - 25 feet above the ground level, so that it can turn round in all the directions and discharge shots at regular intervals. One such device can cover an area of 2.5 - 5.0 acres. Bird scarers are light to carry and are available in many designs to suit the requirements. Because the birds get accustomed to the sound after some days, the equipment has to be shifted to different places often for effective operation.
5. **Flame thrower**. The flame thrower works on the same principle as a kerosene oil pressure stove. It consists of two separate units viz., the pump unit, which is similar to a pneumatic sprayer in which kerosene oil is kept in the fluid tank. The burner unit works on the same principle as that of an Etna burner. The two units are connected by a long rubber tube and a long metal tube through which kerosene oil under pressure is forced into the burner unit. When the kerosene oil is lighted, it vaporizes, mixes with the air and burns. The fire is thrown out through a nozzle at high temperature. This equipment is used to destroy desert locusts, which multiply in large numbers and move in swarms. It is also employed to destroy swarms of caterpillar pests, weed plants etc. **(Fig. 22)**

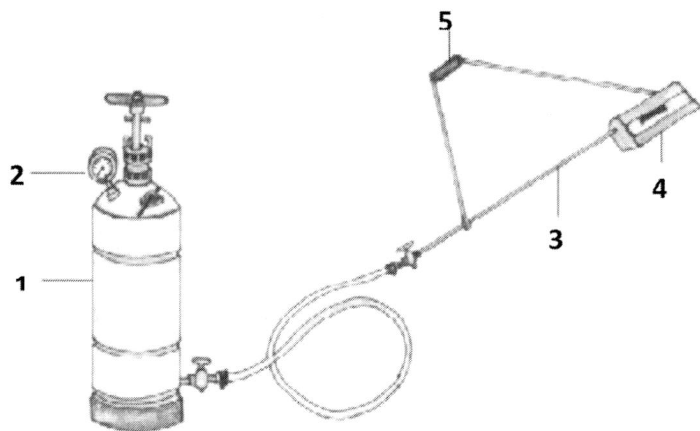


**Fig. 22.** Flame thrower
1. Pneumatic sprayer 2. Pressure gauge 3. Delivery tube for kerosene oil 4. Burner 5. Handle

6. **Rat traps**. Several types of rat traps are used to catch and kill rats in fields, houses, godowns, warehouses, shops etc. Box type of traps made of wood and iron rods, cage type traps made of steel wires, jaw type metal spring traps, Thanjavur bow traps, mud pot traps and many other indigenous types of traps and contraptions are in use. The Thanjavur bamboo bow traps are used on a large scale to destroy field rates in the Thanjavur district and 12 such traps are required to cover one acre of rice field.

# Appendices

**Appendix-1.** Economic classification of insects

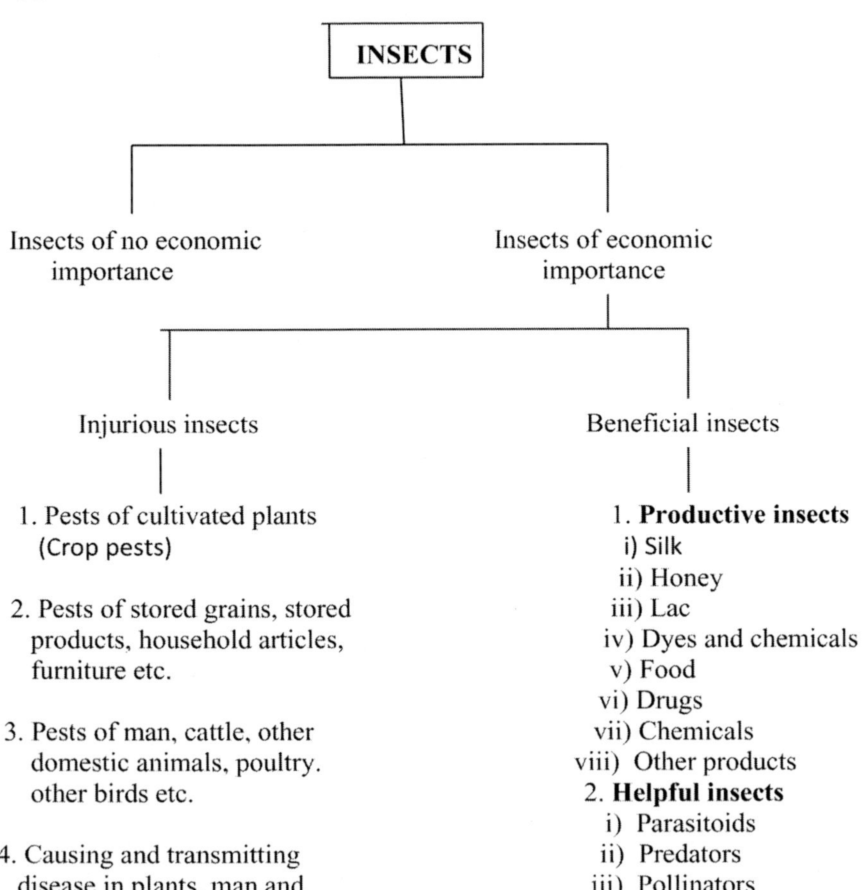

**Appendix-2.** Economic threshold level for some of the important crop pests

| Crop and growth stage | Insect pest | Economic threshold level |
|---|---|---|
| **Rice** | | |
| Vegetative stage | Stem borer | 10% dead hearts or 2 egg masses/m2 or 5 moths/acre in light trap |
| Flowering stage | Stem borer | 2% white ears |
| Vegetative stage | Leaf folder | 50% dead hearts or 2 egg masses/m2 or 5 moths/acre in light trap |
| Flowering stage | Leaf folder | 5 - 6% infested leaves |
| Vegetative stage | Case worm | 25% infested leaves |
| Vegetative stage | Gall fly | 10% galls |
| Vegetative & flowering stage | Brown plant hopper | 1 insect / tiller |
| Vegetative stage | Green leaf hopper | 5 insects/hill or 60 insects/25 net sweeps or 1 insect / hill in RTV endemic areas |
| Flowering stage | Green leaf hopper | 10 insects / hill |
| Vegetative stage | Thrips | 26 insects / 5 hand sweeps |
| Flowering stage | Earhead bug | 5 insects / 100 earheads |
| Milk stage | Earhead bug | 16 insects / 100 earheads |
| **Sorghum** | | |
| Young stage | Shoot fly | 10% deadhearts or 5 adult flies / acre in fish meal trap |
| Vegetative stage | Stem borer | 10% deadhearts |
| **Finger millet** | | |
| Vegetative stage | Stem borer | 10% deadhearts |
| Earhead stage | Earhead bug | 10 insects / earhead |
| **Groundnut** | | |
| Vegetative stage | White flies | 5 - 10 insects / eaf or 5 insects / acre in yellow sticky trap |
| Vegetative stage | Aphids | 15 - 20% infested plants |
| Vegetative stage | Thrips | 1 - 2 insects / leaf |
| Vegetative stage | Stem weevil | 10% galled plants |
| **Crop and growth Cotton** | | |
| Early and late stages | Spotted boll worm | 10% infested shoots or 10% infested bolls |
| Early and late stages | Pink boll worm | 5 moths / acre in pheromone trap or 10% infested shoots or 10% infested flowers or 10% infested bolls |
| Early and late stages | American boll worm | 10% infested bolls or 5 moths / acre in Pheromone trap. |
| **Gram** | | |
| Fruiting stage | Pod borer | 10% infested pods |
| **Field bean** | | |
| Fruiting stage | Pod borer | 10% infested pods |

**Appendix-3.** Important egg parasitoids and their hosts

| Parasitoid | | Host (Eggs) |
|---|---|---|
| **Genus- Trichogramma** | | |
| 1. | *Trichchogramma chilonis* | Sugarcane internode borer, cotton American boll worm, cotton spotted boll worm, cotton pink boll worm, rice leaf folder, tomato fruit borer. |
| 2. | *T. minutum* | Sugarcane early shoot bore, sugarcane internode borer, sugarcane top borer. |
| 3. | *T. evanescens* | Rice stem borer, cotton spotted boll worm, citrus butterfly caterpillar. |
| 4. | *T. japonicum* | Rice stem borer |
| 5. | *T. natum* | Rice stem borer |
| 6. | *T.intermedium* | Sugarcane early shoot borer, sugarcane top borer. |
| 7. | *T. australicum* | Castor semilooper |
| 8. | *T. achaeae* | Castor semilooper |
| 9. | *T. confusum* | Castor semilooper |
| **Genus - *Telenomus*** | | |
| 10. | *Telenomus flanderri* | Castor semilooper |
| 11. | *T. manolus* | Groundnut red hairy caterpillar |
| 12. | *T. beneficiens* | Rice stem borer, sugarcane early shoot borer, sugarcane top borer. |
| 13. | *T. dignus* | Rice stem borer, sugarcane top borer |
| 14. | *T. rovani* | Sugarcane top borer |
| **Family - Eulophidae, Genus -*Tetrastichus*** | | |
| 15. | *Tetrastichus schoenobii* | Rice stem borer, sugarcane top borer. |
| 16. | *T. pyrillae* | Sugarcane leaf hopper |

## **Appendix-4.** Important larval parasitoids and their hosts

| **Parasitoids Hosts (Larva)** | | |
|---|---|---|
| Order - Hymenoptera, Family - Braconidae | | |
| Genus - *Bracon* | | |
| 1. | *Bracon kitcheneri* | American cotton boll worm, cotton pink boll worm, gingelly shoot and leaf webber |
| 2. | *B. brevicornis* | Coconut black-headed caterpillar |
| 3. | *B. greeni* | Egg plant fruit and shoo borer |
| 4. | *B. albolineatus* | Sugarcane early shoot borer, sorghum stem borer |
| 5. | *B. hebator* | Cotton boll worms |
| 6. | *B. chinensis* | Sorghum stem borer |
| 7. | *B.lefroyi* | Cotton boll worms |
| | *Genus - Microbracon* | |
| 8. | *Microbraconchilocida* | Sorghum stem borer |
| 9. | *M. serinopae* | Coconut black-headed caterpillar |
| 10. | *M. chilonis* | Sugarcane early shoot borer |
| 11 | *M. lefroyi* | Field bean pod borer |
| | *Genus - Chelonus* | |
| 12. | *Chelonus narayani* | Sorghum stem borer, Cotton American boll worm |
| | *Genus - A panteles* | |
| 13. | *Apanteles ruficrus* | Rice stem borer |
| 14. | *A. schoenobii* | Rice stem borer |
| 15. | *A. colemani* | Sorghum stem borer |
| Family - Bethyloidae, Genus - *Perisierola* | | |
| 16. | *Perisierola nephantidis* | Coconut black-headed caterpillar |
| | *Family - Ichneumonidae, Genus - Isotima* | |
| 17. | *Isotima javensis* | Rice stem borer, sugarcane top borer |
| | *Genus - Eriborus* | |
| 18. | *Eriborus trochanteratus* | Coconut black-headed caterpillar |
| 19. | *Brachymeria nephantidis* | Larvae of several insects belonging to Order – Lepidoptera , Diptera, coleoptera etc. |
| Family - Chalcididae, Genus - *Brachymeria* | | |
| 20. | *Tetrastichus nyemitawus* | Sorghum shoot fly |
| 21. | *T. compatorensis* | Sorghum shoot fly |
| 22. | *T. ophiusae* | Castor semilooper |
| 23. | *T. ovulorum* | Egg plant lady bird beetle |
| 24. | *T. pyrillae* | Aphids |
| Family - Elasmidae, Genus - *Elasmus.* | | |

| 25. | *Elasmus nephantidis* | Coconut black-headed caterpillar |
|---|---|---|
| | *Order - Diptera, Family - Tachinidae* | |
| | Genus - *Sturmiopsis* | |
| 26. | *Sturmiopsis inferens* | Sugarcane early shoot borer, Sugarcane top borer, Sorghum stem borer |

**Appendix-5.** Important pupal parasitoids and their hosts.

| **Parasitoids** | | **Hosts (Pupae)** |
|---|---|---|
| **Order - Hymenoptera, Family - Eulophidae** | | |
| **Genus - *Tetrastichus*** | | |
| 1. | *Tetrastichus ayyari* | Sugarcane early shoo borer, sugarcane nternode borer, sugarcane top borer, sorghum stem borer. |
| 2. | *T. israli* | Coconut black-headed caterpillar |
| **Genus** - *Trichospilus* | | |
| 3. | *Trichospilus pupivora* | Coconut black-headed caterpillar |
| 4. | *T. diatrae* | Sugarcane internode borer, ragi pink borer. |
| **Family - Ichneuomonidae** | | |
| | **Genus - *Isotima*** | |
| 5. | *Isotima javensis* | Sugarcane early shoot borer, sugarcane internode |
| 6. | *I. dommermani* | Sugarcane top borer |
| **Family - Braconidae** | | |
| **Genus - *Chelonus*** | | |
| 7. | *Chelonus rufus* | Cottton spotted boll worms. |
| **Family - Elasmidae** | | |
| **Genus - *Elasmus*** | | |
| 8. | *Elasmus albopictus* | Rice stem borer |

**Appendix-6.** Important fungal and bacterial parasites and their hosts

| **Parasites** | | **Hosts** |
|---|---|---|
| **Fungal parasites** | | |
| 1. | *Metarrhizium anisopliae* | Sugarcane leaf hopper, Grubs of coconut rhinoceros beetle |
| 2. | *Beauveria bassiana* | Coconut rhinoceros beetles and grubs, |
| 3. | *Aspergillus flavus* | Groundnut red hairy caterpillar, sorghum stem borer, coconut back-headed caterpillar |
| 4. | *Entomophthora coronata* | Aphids attacking cruciferous crops |

| | | |
|---|---|---|
| 5. | *Cephalosporium aphidicola* | Aphids attacking cruciferous crops |
| 6. | *Mucor heimalis* | Sugarcane leaf hopper |
| 7. | *Verticillium lecani* | Rice brown plant hopper |
| 8. | *Hirsutella citriformis* | Rice brown plant hopper, rice green leaf hopper |
| 9. | *Fusarium* spp. | Sugarcane leaf hopper |
| **Bacterial parasites** | | |
| 1. | *Bacillus thuringiensis* | Sugarcane early shoot borer, Cabbage and cauliflower green caterpillar |
| 2. | *B. cerius* | Cotton pink boll worm |
| 3. | *Serretia marcescens* | Groundnut red hairy caterpillar, coconut black-headed caterpillar |

## **Appendix-7.** Differences between parasitoids and predators

| S.No. | Parasitoids | Predators |
|---|---|---|
| 1. | Smaller, usually less stronger and not markedly more intelligent than the host. | Stronger, larger and usually more intelligent than their prey. |
| 2. | Habitat and environment is created and determined by that of the host. | Habitat is independent of its prey |
| 3. | Attack on the host is well planned | Attack on the prey is casual and not well planned. |
| 4. | Lives in or on the body of the host and kills it slowly. | Seizes and devours the prey rapidly. |
| 5. | Usually sluggish once the host is secured. | Very active in habit. |
| 6. | A parasitoid usually completes its life cycle in a single host and so destroys only a single host. | A single predator may attack and feed on several hosts within a short period |
| 7. | Life cycle is short | Life cycle is generally long and may last for several days. |
| 8. | Exhibits marked host specificity in most of the cases and the range of host species is very much limited. | Mostly generalized feeders, except for a few such as, lady bird beetles and hover flies, which inhibit certain specificity to a few hosts. |
| 9. | Exhibits marked host specificity in most of the cases and the range of host species is very much limited. | Mostly generalized feeders, except for a few such as, lady bird beetles and hover flies, which exhibit certain specificity to a few hosts. |
| 10. | Organs of locomotion and mouthparts not very well developed and sometimes even much reduced. | Organs of locomotion, sense organs and mouthparts well developed and may possess special adaptations. |
| 11. | Ovipositor strong, long, well developed and specially adapted for laying eggs on the hosts, which may be inside the stem, leaf etc. | Ovipositor shows no special development |

## Appendix-8

The Ministry of Agriculture and Farmer's Welfare, Government of India has so far banned 46 pesticides and 4 pesticide formulations for import, manufacture or use in the country. In addition 8 pesticide registrations have been withdrawn and 9 pesticides have been placed under restricted use.

**List of Pesticides / Formulations - Banned in India as on 01-10-2022. Pesticides banned for manufacture, Import and Use**

| | | | |
|---|---|---|---|
| 1. | Alachlor | 2. | Aldicarb |
| 3. | Aldrin | 4. | Benzene Hexachloride |
| 5. | Benomyl | 6. | Calcium Cyanide |
| 7. | Carbaryl | 8. | Chlorbenzilate |
| 9. | Chlordane | 10. | Chlorfenvinphos |
| 11. | Copper Aceto Arsenite | 12. | Diazinon |
| 13. | Dibromo Chloropropane (DBCP) | 14. | Dichlorvos |
| 15. | Dieldrin | 16. | Endosulfan |
| 17. | Endrin | 18. | Ethyl Mercury Chloride |
| 19. | Ethyl Parathion | 20. | Ethylene Dibromide (EDB) |
| 21. | Fenarimol | 22. | Fenthion |
| 23. | Heptachlor | 24. | Lindane (Gamma HCH) |
| 25. | Linuron | 26. | Maleic Hydrozide |
| 27. | Menazon | 28. | Methoxy Ethyl Mercury Chloride |
| 29. | Methyl Parathion | 30. | Metoxuron |
| 31. | Nitrofen | 32. | Paraquat Dimethyl Sulphate |
| 33. | Pentachlorophenol | 34. | Pentachloro Nitrobenzene |
| 35. | Phenyl Mercury Acetate | 36. | Phorate |
| 37. | Phosphamidon | 38. | Sodium Cyanide |
| 39. | Sodium Methane Arsenate | 40. | Tetrodifon |
| 41. | Thiometon | 42. | Toxaphene |
| 43. | Triazophos | 44. | Tridemorph |
| 45. | Trichloro Acetic Acid | 65. | Trichlorfon |

**Pesticide formulations banned for import, manufacture and use**

1. Carbofuran 50% SP
2. Methomyl 12.5% L

3. Methomyl 24% Formulation
4. Phosphamidon 85% SL

**Pesticides withdrawn**

1. Ferbam
2. Formothion
3. Nickel Chloride
4. Paradichloro benzene (PDCB)
5. Simazine
6. Sirmate
7. Warfarin

**Pesticides restricted for use in the country**

1. Aluminium Phosphide - It can be used only by Govt./Govt. undertakings/ Govt. organizations/pest control operators etc. only under strict supervision.

   Production, marketing and use of Aluminium phosphide in tube packs containing 3.0 g. tablets by the public is completely banned.
2. Cypermethrin - Cypermethrin 3% smoke generator is to be used only through Pest Control Operators under strict supervision and not allowed to be used by the General Public.
3. Dazomet - Use of Dazomet is not permitted on Tea crop.
4. Dichloro Diphenyl Trichloro Ethane (DDT) - Use of DDT in Agriculture is withdrawn.
5. Fenitrothion - Use of Fenitrothion is banned in Agriculture, except for locust control and in Public Health Organizations.
6. Methyl Bromide - Methyl Bromide may be used only by Govt. /Govt. undertakings / Govt. Organizations /Pest Control Operators etc. under strict supervision.
7. Monocrotophos - Monocrotophos is banned for use on vegetable crops.

**Appendix-9.** List of insecticides approved by the Government of India for plant protection purposes, their formulations and dosages.

| No. | Chemical name | Trade name | Formulation (% ai) | Dusting (per ac.) | Quantity to be applied | | |
|---|---|---|---|---|---|---|---|
| | | | | | Soil application (per ac.) | Spraying | |
| | | | | | | Active (% ai) | ml./g. per lit. of water |
| (1) | (2) | (3) | (4) | (5) | (6) | (7) | (8) |
| 1. | Acephate | Starthene, Orthene, Lancer, Acemil | 75 WSP | - | - | 0.15 | 2.0g |
| 2. | Acephate | Hunk, Starphate Super | 95 SG | - | - | 0.12 | 1.25g |
| 3. | Acetamiprid | Manik, Ikon, Proud, Baadshah, Dhanpeet | 20 SP | - | - | 0.08 | 0.8g |
| 4. | Alpha Cypermethrin | Alfastar, Guru, Dash, Dolphin, Safari | 10 EC | - | - | 0.005 | 0.125ml |
| 5. | Alphamethrin | Legend, Mig 10, Tata Alpha, Thrill, Alpha, Alfagold. | 10EC | - | - | 0.005 | 0.125ml. |
| 6. | Bifenthrin | Hectastar, Metastar, Markar, Super Star, Canister, Cntrix. | 10 EC | - | - | 0.005 | 0.125ml. |
| 7. | Buprofezin | Jawaa, Flotis, Trust, Devifezin, Applaud. | 25 EC | - | 13 kg. | - | - |
| 8. | Carbofuran | Furadan, Carbogran, Vinfuran, Curater. | 3 G | - | - | 0.005 | 2.0ml. |

| | | | | | | | |
|---|---|---|---|---|---|---|---|
| 9. | Carbosulfan | Marshal, Aatank, Carbosulfan | 25 EC | - | - | 0.005 | 2.0 ml. |
| 10. | Cartap | Padan, Thiobel, Vegetox | 50 WSP | - | - | 0.10 | 2.0 g |
| 11. | Cartap | Caldan, Katsu, Marktap, Indi Gold | 4 G | - | 10 kg | - | - |
| 12. | Chlorantraniliprole | Kashima, Vistara, Ferterra, Stambh | 0.4 G | - | 6.0 kg. | - | - |
| 13. | Chlorantraniliprole | Shimo, Cover Coragen. | 18.5 SC | - | - | 0.005 | 3.0 ml. |
| 14. | Chlorfenapyr | Intreprid, Record, Lepido, Cutlass | 10 SC | - | - | 0.005 | 0.0125 ml. |
| 15. | Chlorpyriphos | Meesho, Hindol, Chloro, Capban | 1.5 D | 10 kg | - | - | - |
| 16. | Chlorpyriphos | Krishan, Starban, Deviban, Megaban | 20 EC | - | - | 0.04 | 2.0 ml. |
| 17. | Chlorpyriphos | Terminator, Clorocon Chloroxa-50 | 50 EC | - | - | 0.075 | 1.5 ml. |
| 18. | Chlorpyriphos | Deviban 10G, Hilban, Suldrin | 10 G | - | 10 kg | - | - |
| 19. | Chromafenozide | Dodger, India Mart Matrix. | 80 WP | - | - | 0.02 | 0.25 ml. |
| 20. | Clothianidin | Dantotsu | 50 WDG | - | - | 0.15 | 0.03g |
| 21. | Cypermethrin | Ripcord, Cypermar, Challenge, Lacer | 10 EC | - | - | 0.005 | 0.5 ml. |
| 22. | Cypermethrin | Cymbush, Cypervip Sandoz Cyperkill | 25 EC | - | - | 0.0075 | 0.3 ml. |
| 23. | Deltamethrin | Decis, Delta Hit, Redox, Ecdel 28 | 2.8 EC | - | - | 0.28 | 1.0 ml |

| | | | | | | | |
|---|---|---|---|---|---|---|---|
| 24. | Deltamethrin | Derin, Delta Super, Delta 11, Tokiton. | 11.0 EC | - | - | 0.33 | 0.3 ml |
| 25. | Diafenthiuron | Derby, Ashwamed, Shoku, Pager. | 50 WP | - | - | 0.05 | 1.25 g |
| 26. | Dimetoate | Rogor, Cygon, Champ, Sagar. | 30 EC | - | - | 0.03 | 1.0 ml. |
| 27. | Dicofol | Kelthane, Acarin, Micothane, Hexakel | 18.5 EC | - | - | 0.37 | 0.6 g. |
| 28. | Dinotefuran | Osheeen, Dinotrex, Oshin, Token | 20 SG | - | - | 0.12 | 0.6 g. |
| 29. | Emamectin | Commander, Egro, Jupiter, Ema Gold | 5.0 SG | - | - | 0.005 | 1.0 g. |
| 30. | Etofenprox | Dash, Etofenprox, Trebon | 30 EC | - | - | 0.15 | 0.5 ml. |
| 31. | Fenobucarb | Wardon, Lawin, Pinaka, Pestanal | 50 EC | - | - | 0.75 | 0.66 ml. |
| 32. | Fenpropathrin | Danitol | 10 EC | - | - | 0.2 | 2.0 ml. |
| 33. | Fenpropathrin | Meothrin | 30 EC | - | - | 0.15 | 0.5 ml. |
| 34. | Fenpyroximate | Pyromite, Akari, Seagate, Sadna. | 5.0 EC | - | - | 0.005 | 1.0 ml. |
| 35. | Fenvalerate | Fielder, Anufen, Fnoxx, Ghatak. | 0.4 DP | **10 kg.** | - | - | - |
| 36. | Fenvalerate | Fenhit, Sumicidin, Tatafen, Fenval. | 20 EC | - | - | 0.02 | 1.0 ml. |
| 37. | Fipronil | Recent, Katyayani, Jaanbaaz, Fipro | 0.3 G | - | 10 kg. | - | - |
| 38. | Fipronil | Chilli KOR, Agenda, Bio M Power. | 2.92 EC. | - | - | 0.01 | 2.0 ml. |

| | | | | | | | |
|---|---|---|---|---|---|---|---|
| 39. | Fipronil | Aashirwaad, Bell, Recelt, Getter. | 5.0 SC | - | - | 0.001 | 0.20 ml. |
| 40. | Fipronil | Fipro 80, Jump Joker, Ahead | 80.0 WG | - | - | 0.001 | 0.12 ml. |
| 41. | Flubendiamide | Flame, Fluben, Superzite, Flubensik | 39.37 SC | - | - | 0.10 | 0.5 ml. |
| 42. | Flubendiamide | Takibi, Tata Takumi | 20 WG | - | - | 0.005 | 1.0 ml. |
| 43. | Hexythiazox | Cubax, Maiden, Miteban, Mitolin. | 5.45 EC | - | - | 0.005 | 1.0 ml. |
| 44. | Imidachloprid | Green mida, Media, Confidor, IMD-178 | 17.8 SL | - | - | 0.008 | 0.5 ml. |
| 45. | Imidachloprid | Green mida, Isogashi, Tata mida, Hi-amida | 30.5 SC | - | - | 0.75 | 0.25 ml. |
| 46. | Imidachloprid | Admit, Admire, IMD-70, Ad-fyre. | 70 WG | - | - | 0.15 | 0.75 ml. |
| 47. | Imidachloprid | Aakrosh, Farmida, Gaucho, Romix. | 48 SL/FS (For seed treatment at 0.5- 0.9 ml. per kg . of seeds) | - | - | - | - |
| 48. | Indoxacarb | King Doxa, King Carb, K Indox | 14.5 SC | - | - | 0.15 | 1.0 ml. |
| 49. | Lambda Cyhalothrin | Kozuka, Reeva, Karate, Lamdex. | 2.5 EC | - | - | 0.006 | 2.5 ml. |
| 50. | Lambda Cyhalothrin | Metador, Jackpot, Attack, Mantra, Killer. | 4.9 CS | - | - | 0.007 | 1.5 ml. |
| 51. | Lambda Cyhalothrin | Lamdex Gold, Neon, Clamda-5, Singham. | 5.0 EC | - | - | 0.007 | 1.5 ml. |
| 52. | Malathion | Growhit, K thion, apalon | 5.0 DP | **10 kg.** | - | - | - |

| | | | | | | | |
|---|---|---|---|---|---|---|---|
| 53. | Malathion | Cythion, Suthion, Mal-50, Malathion. | 50 EC | - | - | 0.65 | 1.25 ml. |
| 54. | Methomil | Dash, Kinet, Dragon, Atom-40, Lannate. | 40 SP | - | - | 0.60 | 1.5 ml. |
| 55. | Methyl Parathion | Missile, Paradol, Gramexin, Tagpar. | 2.0 DP | **10 kg.** | - | - | - |
| 56. | Methyl Parathion | Folidon, Kildot 50E, Devithion 50. | 50 EC | - | - | 0.5 | 1.5 ml. |
| 57. | Monocrotophos | Monster, Monorin, Monogreen, Monophos. | 36 SL | - | - | 0.45 | 1.25 ml. |
| 58. | Novaluron | Pedestol, Rimon, Novamax, Noval. | 10 Ec | - | - | 0.02 | 2.0 ml. |
| 59. | Permethrin | Signor, Pixel-25, Proud, Brahmastra | 25 EC | - | - | 0.05 | 2.0 ml. |
| 60. | Phenthoate | Phendal, Pestanal, Royal 50, Cidial 50. | 50 EC | - | - | 0.10 | 2.0 ml. |
| 61. | Profenophos | Profen, Promax, Pro kil, Pestanal | 50 EC | - | - | 0.75 | 1.5 ml. |
| 62. | Propergite | Omite, Simbaa, Mite Kill, Teeka. | 57 EC | - | - | 0.10 | 2.0 ml |
| 63. | Pymetrozine | Chess, Apply, Inbox, Suruga, Metapro. | 50 WG | - | - | 0.037 | 0.75 ml. |
| 64. | Quinalphos | Molquin, Phoschem, Quinalphos 1.5% DP. | 1.5 DP | **10 kg.** | - | - | - |
| 65. | Quinalphos | Krush, Ekalux, Goldlux, Dhanulux. | 25 EC | - | - | 0.125 | 2.5 ml. |
| 66. | Spinosad | Spino 25, Success, Spinosad 2.5. | 2.5 EC | - | - | 0.003 | 1.5 ml |

| 67. | Spinosad | Natroba, Taffin, Tracer, Spinter. | 45 SC | - | - | 0.18 | 0.4 ml. |
|---|---|---|---|---|---|---|---|
| 68. | Spiromesifen | Voltage, Chemipider, Offmite, Oberon. | 22.9 SC | - | - | 0.23 | 1.0 ml. |
| 69. | Thiachloprid | Alanto, Gunwaan, Splendour, Scil. | 21.7 SC | - | - | 0.10 | 0.4 ml. |
| 70. | Thiodicarb | Larvin, Spiro | 75 WP | - | - | 0.02 | 0.3 ml. |
| 71. | Thiamethoxam | Anant, Theme, Actara, Evident, Pestanal. | 25 WG | - | - | 0.10 | 0.4 ml. |
| 72. | Thiamethoxam | Capsadis, Devasena, Shutter, Thioxam Super | 75 SG | - | - | 0.02 | 0.3 ml. |
| 73. | Triazophos | Shooter, Jane, Triazophos 20. | 20 EC | - | - | 0.05 | 2.5 ml. |
| 74. | Triazophos | Shook, Falsop, Rider, Growthion. | 40 EC | - | - | 0.05 | 1.25 ml. |

**Appendix-10.** Types of different Plant Protection chemical formulations

| | |
|---|---|
| DP | Dustable powders |
| DS | Powder for dry seed treatment |
| DT | Tablets for direct application |
| EG | Emulsifiable granules |
| EP | Emulsifiable powders |
| GR | Granules |
| GW | Water soluble gel |
| SG | Water soluble granules |
| SP | Water soluble powder |
| SS | Water soluble powder for seed treatment |
| ST | Water soluble tablet |
| WG | Water dispersible granules |
| WP | Wettable powders |
| WS | Water dispersible powders for slurry seed treatment |
| WT | Water dispersible tablet |
| LS | Solution for seed treatment |
| SL | Soluble concentrate |
| UL | Ultra low volume (ULV) liquids |
| DC | Dispersible concentrates |
| EC | Emulsifiable concentrates |
| ES | Emulsions for seed treatment |
| CS | Capsule suspension |
| FS | Suspension concentrate for seed treatment |
| OB | Oil-based suspension concentrates |
| SC | Suspension concentrates |

**Appendix-11.** Quantity of spray fluid that may be required to spray one acre of crop by different types of sprayers.

| **Crop** | **High volume Sprayer** | **Semi low volume Sprayer** | **Low volume Sprayer** |
|---|---|---|---|
| Rice | 200 - 250 lit. | 50 - 100 lit. | 5.0 - 50 lit. |
| Groundnut | 250 - 300 lit. | 65 - 120 lit. | 6.0 - 60 lit. |
| Cotton | 300 - 500 lit. | 75 - 200 lit. | 7.5 - 100 lit. |
| Millets | 200 - 350 lit. | 50 - 140 lit. | 5.0 - 70 lit. |
| Potato | 300 - 350 lit. | 75 - 140 lit. | 7.5 - 70 lit. |
| Pulse crops | 250 - 350 lit. | 65 - 140 lit. | 6.0 - 70 lit. |
| Sugarcane | 350 - 500 lit. | 85 - 200 lit. | 9.0 - 100 lit. |
| Wheat | 250 - 350 lit. | 65 - 140 lit. | 6.0 - 70 lit. |

| Chillies | 200 - 350 lit. | 50 - 140 lit. | 5.0 - 70 lit. |
|---|---|---|---|
| Tomato | 200 - 350 lit. | 50 - 140 lit. | 5.0 - 70 lit. |
| Egg plant | 200 - 350 lit. | 50 - 140 lit. | 5.0 - 70 lit. |
| Grapevine | 500 - 600 lit. | 125 - 240 lit. | 13 - 120 lit. |
| Banana | 250 - 600 lit. | 65 - 240 lit. | 6.0 - 120 lit. |
| Coconut | 3,0 lit. / tree | 0.75 lit. / tree | 0.5 lit. / tree |
| Mango | 5.0 - 10 lit. / tree | 1 25 - 4.0 lit. /tree | 0.125- 2.0 lit. / tree |
| Guava | 5.0 - 10 lit. / tree | 1.25 - 4.0 lit. /tree | 0.125 - 2.0 lit. / tree |
| Sapota | 5.0 - 10 lit. / tree | 1.25 - 4.0 lit. /tree | 0.125 - 2.0 lit../tree |
| Citrus plants | 5.0 - 10 lit. / tree | 1.26 – 4.0 lit. /tree | 0.125 - 2.0 lit../tree |

While using semi - low volume or low volume sprayers, the quantity of chemical formulation , which is required to spray the crop with a high volume sprayer should be used, although the quantity of water required to prepare the spray fluid is much less. Irrespective of the quantity of water used for preparing the spray fluid, the chemical formulation to be used should be the same. Because a lesser quantity of water is used for preparing the spray fluid in the case of semi - low and low volume sprayers, the formulations should not be reduced proportionately.

**Appendix-12.** Ready reckoner to find out the quantity of various insecticide formulations required to prepare de-sired concentrations of spray fluid (ml. / g. for every 100 litres of water)

| **Active ngredient in the spray fluid (%)** | **Active ingredient in the commercial insecticide formulation (%)** | | | | | | | | | | |
|---|---|---|---|---|---|---|---|---|---|---|---|
| | 6.5 | 18 | 20 | 25 | 30 | 35 | 40 | 50 | 75 | 85 | 100 |
| 0.005 | 76.9 | 27 8 | 25 | 20 | 17 | 14.3 | 12.5 | 10 | 6.7 | 6.0 | 5 |
| 0.01 | 153.8 | 55.6 | 50 | 40 | 34 | 28.6 | 25.0 | 20 | 13.3 | 12.0 | 10 |
| 0.015 | 230.7 | 83.4 | 75 | 60 | 51 | 42.9 | 37.5 | 30 | 20.0 | 18.0 | 15 |
| 0.02 | 307.6 | 111.2 | 100 | 80 | 68 | 57.2 | 50.0 | 40 | 26.5 | 24.0 | 20 |
| 0.025 | 384.5 | 139.0 | 125 | 100 | 85 | 71.5 | 62.5 | 50 | 33.5 | 29.0 | 25 |
| 0.03 | 461.4 | 166.8 | 150 | 120 | 102 | 85.8 | 75.0 | 60 | 40.0 | 35.0 | 30 |
| 0.035 | 538.3 | 194.6 | 175 | 140 | 119 | 100.0 | 87.5 | 70 | 46.6 | 41.0 | 35 |
| 0.04 | 615.2 | 222.4 | 200 | 160 | 136 | 114.4 | 100.0 | 80 | 53.0 | 47.0 | 40 |
| 0.045 | 692.1 | 280.2 | 225 | 180 | 153 | 128.7 | 112.5 | 90 | 59.8 | 53.0 | 45 |
| 0.05 | 769.0 | 278.0 | 250 | 200 | 167 | 143.0 | 125.0 | 100 | 66.4 | 59.0 | 50 |
| 0.075 | 1153.5 | 417.0 | 375 | 300 | 252 | 214.4 | 187.5 | 150 | 100.0 | 88.0 | 75 |
| 0.1 | 1538.0 | 556.0 | 500 | 400 | 334 | 286.0 | 250.0 | 200 | 133.3 | 118.0 | 100 |
| 0.2 | 3076.0 | 1112.0 | 1000 | 800 | 668 | 572.0 | 500.0 | 400 | 266.6 | 235.0 | 200 |
| 0.25 | 3845.0 | 1390.0 | 1250 | 1000 | 836 | 715.0 | 825.0 | 500 | 3 33.1 | 294.0 | 250 |
| 0.3 | 4614.0 | 1668.0 | 1500 | 1200 | 1002 | 858.0 | 750.0 | 600 | 399.9 | 353.0 | 300 |
| 0.5 | 7690.0 | 2780.0 | 2500 | 2000 | 1679 | 1430.0 | 1250.0 | 1000 | 666.5 | 588.0 | 500 |
| 0.75 | 11535.0 | 4170.0 | 3750 | 3000 | 2505 | 2145.0 | 1875.0 | 1500 | 999.3 | 882.0 | 750 |
| 1.0 | 15380.0 | 5560.0 | 5000 | 4000 | 3340 | 2860.0 | 2500.0 | 2000 | 1333.0 | 1176.0 | 1000 |

**Appendix-13.** Classification of pesticides on acute toxicity level

| Particulars | Category I Extremely toxic | Category II Highly toxic | Category III Moderately toxic | Category IV Slightly toxic |
|---|---|---|---|---|
| $LD_{50}$ | | | | |
| Oral (mg./kg.) | 0.1 - 50 | 51 - 500 | 501 - 5,000 | > 5,000 |
| Dermal (mg./kg.) | 0.1 - 200 | 201 - 2,000 | 2,001 - 20,000 | >20,000 |
| Color of the lower triangle on the label of the container | Red | Yellow | Blue | Green |
| Signal in the upper Triangle | Poison; 'Skull and cross bones' | 'Poison' | 'Danger' | 'Caution' |

**Appendix-14.** Toxicity levels of different synthetic pesticides (mg. / kg. body weight of the test animal)

| Pesticide | Acute oral toxicity for rats (mg./kg.) | Dermal toxicity for rats (mg./kg.) |
|---|---|---|
| Acephate | 700 mg | > 2,000 mg (for rabbit) |
| Acetamiprid | 140 - 417 mg | > 2,000 mg (for rabbit) |
| Alphamethrin | 79 - 400 mg | > 2,000 mg (for rabbit) |
| Carbaryl | 500 - 700 mg | - |
| Carbosulfan | 90 - 250 mg | > 2,000 mg (for rats) |
| Cartap | 325 - 390 mg | > 1,000 mg (for mice) |
| Chlorfenvinphos | 9.6 - 39 mg | - |
| Chlorpyriphos | 135 - 163 mg | - |
| Cypermethrin | 200 - 800 m | - |
| Deltamethrin | 52 mg | > 2,000 mg (for rabbit) |
| Dicofol | 750 mg | > 2,500 mg (for rabbit) |
| Dieldrin | 46 mg | 60 - 90 mg |
| Dimethoate | 245 mg | 700 - 1150 mg |
| Dinocap | 980 - 1190 mg | - |
| Ethion | 96 mg | 245 mg (for rabbit) |
| Fenitrothion | 570 - 800 mg | > 3,000 mg (for mice) |
| Fenthion | 178 - 319 mg | 320 - 330 mg |
| Fenvalerate | 300 - 630 mg | - |
| Formothion | 365 - 500 mg | 400 - 1680 mg |
| Imidachloprid | 450 mg | > 5,000 mg. (for rats) |
| Lindane | 125 mg | - |
| Malathion | 1500 mg | 4,100 mg (for rabbit) |
| Methomil | 17 - 24 mg | > 2,000 mg )for rabbit) |

| Methyl demeton | 50 mg | 130 mg |
|---|---|---|
| Methyl parathion | 9 - 25 mg | 67 mg |
| Monocrotophos | 21 mg | 354 mg (for rats) |
| Permethrin | 430 - > 4000 mg | - |
| Phenthoate | 300 - 400 mg | 700 - 1400 mg |
| Phosalone | 100 - 180 mg | 1500 mg |
| Phosphamidon | 17.9 - 30 mg | 374 - 530 mg |
| Propoxur | 90 - 128 mg | 200 - 1000 mg |
| Profenofos | 358 mg | 3300 mg |
| Propetemphos | 59.5 - 119 mg | - |
| Quinalphos | 62 - 137 mg | 800 – 1400 mg |
| Thiodicarb | 66 - 120 mg | > 2,000 mg (for rabbit) |
| Thiometan | 85 - 225 mg | 300 mg |
| Triazophos | 57 - 68 mg | > 2000 mg |
| Trichlorphon | 450 - 500 mg | > 2000 mg |
| Wettable sulphur | non - toxic | Causes eye irritation |

# References

Hem Singh Pruthi. 1969. Textbook on Agricultural Entomology. Indian Council of Aricultural Research, New Delhi. Pp.977.

Kanna, S.S. 1983. Agricultural Entomology. Adarsh Prakashan, Agra. pp.304.

Krishnan, N.T. 1993. Economic Entomology. J. J. Publications, Madurak. Pp.265.

Lewin Devasahayam, H. 2011. Practical Manual of Entomology - Insect and non-insect pests. New India Publishing Agency, New Delhi. pp.392

Pren Mohan Nigam and Ashok Kumar.1991. Plant Protection - Insect control. Emkay Publications, Delhi. pp. 191.

# Glossary of Scientific Terms

**A**

**Acaricide** - A chemical or chemical formulation employed to control to control mites and ticks; alternate name 'miticide

**Acquired immunity** - Resistance to a microbial or other antigenic substance acquired either actively or passively by a naturally susceptible individual.

**Active ingredient (a.**i.) - Toxic component present in a formulated pesticide.

**Aerosol** - A pesticide formulation that is dispersed in the form of spray droplets of diameter 0.1-5.0 µm. with the help of compressed air, usually from a canister.

**Aflatoxins** - Organic compounds or metabolites produced by the fungus *Aspergillu flavus*, which are highly toxic and carcinogenic to mammals.

**Aldicarb** - A synthetic, systemic carbamate insecticide , acaricide and nematicide

**Alternate hosts** - Host plants belonging to Genera other than the genus of the primary host; unlike hosts; host plants other than the primary host on which a microorganism (fungus, bacterium, virus etc.) develops to complete its life cycle. Weeds and wild plants are often alternate hosts of certain pests and diseases.

**Aluminium phosphide** - A chemical formulation commercially available as tablets that is used as a fumigant for he control of storage pests.

**Amino acids** - The basic organic compounds that contain the amino group and the carboxyl group required for the synthesis of protein.

**Anthocyanine** - Sap soluble glycosides imparting scarlet, purple and blue colors in insects.

**Antidote** - Antidote is a drugthat counteracts (neutralizes) the effect of another drug or poison.

**Antifeedant** - A natural or synthetic chemical substance, which acts either to inhibit the stimulation of gustatory receptors that normally recognize suitable food or to stimulate receptors, which elicit a negative response to food.

**Atrophy** - Arrrested development or loss of a part or organ incidental to the normal development or life of an animal or plant

**Attractants** - Materials that are able to attract insects and make them eat or imbibe the poison used in poison baits and consequently cause their death.

**B**

**Bait** - Foodstuff used for attracting pests, which are usually mixed with a poison to form poison bait; an edible material that is attractive to the pest, which contains a pesticide unless used as a prebait.

**Basal treatment** - A treatment applied to the stems or trunks of plants at and just above the ground level

**Biological control** - Manipulation of natural enemies such as, parasitoids, predators or other microorganisms in an attempt at reducing pest population and keep them at much reduced level; total or partial destruction of pest population by other organisms.

**Black light traps** - In black light traps 'black light' or 'Ulytaviolet light' is used for collecting many insects that are active and flying at night and are attracted to light.

**Bot** - The larvae of certain flies that are parasitic in the body of mammnals.

**Botanical insecticide** - An insecticide produced from a plant or plant product such as, Pyrethrum, azadirachtin etc.

**Broadcast application** - Application of insecticide granules by hand or by applicators over a surface area.

**Broad spectrum insecticide** - Insecticides effective against a variety of pests or a broad range of targets usually acting generally on the insect nervous system; non-selective, having almost the same toxicity to a wide range of insect pests.

**C**

**Chlorophyll** - The green photosynthetic pigment found in the chloroplasts of plants, which converts sunlight energy into chemical energy.

**D**

**Dead heart** - Wilting and eventual death of the central shoot in the case of monocots, which come off quite easily when pulled; Wilting and death of the main and side shoots in the case of dicots, as young caterpillar bores into the stem of the plant either from the top or side, eats the inner tissue and makes tunnels inside (See white ears)

**Defoliator** - Any biting and chewing insect that destroys the leaves of plants extensively.

**Deterrent** - A chemical substance that deters feeding or oviposition of an insect.

**Detritivorous** - Feeding on organic debris of plant or animal origin.

**E**

**Ecdysone** - Ecdysone secreted by the prothoracic ecdysal glands of immature insects is the major polyhydroxy steroid hormone in insects that plays an essential role in coordinating developmental transitions such as, larval moulting and metamorphosis.

**Economic control** - Control of a pest in which the expenditure on control measures is more than compensated by increased value of yield.

**Economic injury level (EIL)** - The infestation level of a particular pest at which the damage caused by the pest will result in economic loss; the lowest population density of a pest that causes economic damage.

**Economic pest** - A pest causing a crop loss of 5.0 - 10.0 per cent.

**Economic threshold level (ETL)** - The infestation level of a particular pest, beyond which level there will be economic loss; pest population level at which control measures have to be initiated to prevent the pest population rising to the economic injury level.

**Ecosystem** - A self containing habitat in which the living organisms and the physiochemical environment interact in an exchange of energy and mater to form a continuing cycle.

**Ectoparasie** - A parasite that lives externally and feeds from the host by sending in specialized organs or mouthparts to ingest nourishment from the host, but does not kill the host; a parasite feeding on a host from the exterior.

**Egg parasitoids** - The parasitoids deposit their eggs inside or outside the host eggs and the progeny of the parasitoid emerging from the eggs, parasitize the host eggs and destroy them (e.g.) *Trichogramma* spp. on Lepidopteran eggs.

**Emulsifiable concentrate (EC)** - An insecticide formulation in which the active ingredient is dissolved in a solvent and when the formulation is mixed with water, the active ingredient is held in a suspended state in water.

**Emulsifier** - In an emulsion, the active ingredient is dissolved in an oil-based solvent. An emulsifier allows the active ingredient and the solvent to mix evenly with water before application.

**Endemic pest** - A pest permanently established in a moderate or severe form in a defined area, commonly a country or part of a country.

**Endoparasite** - A parasite, which enters a host, lives within the host and feeds from the host; a parasite, which enters a host and feeds from within.

**Entomophagous-** Organisms, which feed on insects or their parts; insecivorous.

**Epidemic (Epiphytotic)** - A sudden and rapid flare up of a pest from its endemic area; A widespread and severe, but temporary flare up in the incidence of a pest, particularly within a season.

**Epidemiology** - The study of pest occurrence and spread throughout a host population.

**Exotic pest** - Pest introduced from a foreign country; not native.

**F**

**Field resistance** - The resistance observed under field conditions due to the exposure of plants to natural population of insects; Moderate resistance.

**Frond** - The entire leaf of a coconut palm; The fleshy mid rib of the frond consists of 100 - 120 leaflets on either side.

**Fumigant** - A substance or mixture of substances, which produce gas vapor, fume smoke intended to destroy insects or other pests (e.g.) Carbon disulphide, carbon tetrachloride, chloropicrin, dibromo chloropropane, EDCT, ethylene dichloride , hydrogen cyanide, methyl bromide, phostoxin etc.

**Fumigation** - The application of a fumigant for disinfestation of an area.

**Furrow application** - Placement of pesticide with seed or seed material in the furrow at the time of sowing.

**G**

**Gall** - A swelling or outgrowth produced in a plant as a result of attack by an insect, mite or any microorganism.

**Gastric lavage** - Commonly called gastric irrigation, it is the process of cleaning out the contents of the stomach using a tube and it is one of the most routine means of eliminating poison from the stomach.

**Gelatin capsules** - Capsules filled with pesticide for use as root, stem or zone application.

**Granulosis virus (GV)** - Viruses causing insect diseases characterized by the presence of large number of very small, but microscopically dissembled granular inclusion bodies in infected cells.

**H**

**Haematophagus** - Blood suckers; Feeding or subsisting on blood.

**Hibernation** - Insects that remain inactive during the winter months undergo a state in which their growth, development and activities are suspended temporarily with a metabolic rate that is high enough to keep them alive. The dormant conmdition is termed diapause.

**Horizontal resistance -** The type of resistance, which is effective against all the known biotypes of the insect; It is quantitative, as the degree of resistance depends on the number of minor genes each contributing a small effect. It is also called non-specific resistance.

**Host plant -** Living plant attacked by or harboring a pest or parasite and from which the invader obtains part or entire requirement of of its nourishment.

**Host range -** The various kinds of host plants that may be attacked by a pest.

**Hydraulic sprayer -** In this type of sprayers the spray fluid is pressurized directly by using an air pump to build up the air pressure above the spray fluid in the air tight container. The pressurized fluid is then forced through the spray lance.

**Hyperparasitoid -** An insect parasite of another primary parasitoid; Also known as secondary parasitoid.

**I**

**Immune -** Cannot be infested by a given pest.

**Infest -** To occupy and cause injury to either a plant, animal, stored product or soil.

**Insecticide -** A toxic chemical substance or compound employed to kill and control insect pests; Chemicals prepared with express purpose of controlling insect pests, the active ingredients of which may be in the form of gas, liquid or solid and formulated as liquefied gases, smokes, dusts, water dispersible powders, emulsions or solutions.

**Insect pest management -** An ecologically based strategy of maintaining insect pest population below the economic injury level by the use of any or all control techniques that are economically, ecologically and socially acceptable. **Integrated control -** An approach that attempts to use all available methods of control of a pest or of all the pests and diseases of a crop plant for best control results, but with least cost and least damage to the environment.

**Integrated pest management (IPM) -** An approach utilizing all available methods of pest suppression including mechanical, biological, chemical, physical and natural control in a systematic way with he primary goal of safe, effective and economical pest population reduction and to maintain the pest population at a level below that causing economic damage or loss. It may be directed at a single important target pest species or at a pest complex, integrating the individual control measures applied against each, so as not to interfere one with the other.

**K**

**Key pest** - An important major pest species in the pest complex attacking a crop and causing economic damage, with a dominating effect on control practices.

**Knock down effect** - An insecticide that has a power to knock down flying insects rapidly, but may not kill them immediately.

**L**

**Larval parasitoid** - Parasitoid that deposits its eggs in or on the host larvae. The progeny of the parasitoid emerges from the host larvae, parasitizes and destroys the host larvae (e.g.) *Bracon brevicornis, Apanteles flavipes* etc.

**$LD_{50}$** or **Median lethal dose** - A value used in presenting mammalian toxicity, usually oral toxicity expressed in mg. of toxicant per kg. of body weight ; A lethal dose for 50 per cent of the test organisms; The dose of a toxicant producing 50 per cent mortality in a population of test organisms.

**Legal control** - Control of pests through the enactment of legislation that enforces control measures or imposes regulations such as, quarantines to prevent introduction of or spread of pests.

**Lethal** - Fatal or deadly.

**Light trap** - A device for collecting insects, consisting of a light source to attract insects at night ; A device to trap insects at night using a light source.

**Longevity** - Life span of an organism or individual.

**M**

**Major pest** - Insect pest that causes a loss of more than 10 per cent of the produce.

**Man-made pest** - A species, which is a pest only because of human interference with natural control processes that normally regulate its density to non-pest levels. Most commonly generated through pest upsets i.e. the unintentional destruction of natural enemies of a non-pest species with a chemical insecticide or through noncultural practices.

**Mechanical control** - Any mechanical means of pest control or mechanical devices for pest control such as, insect and mammalian traps of various kinds, barriers to prevent entry or gaining access of pests to plants or other commodities.

**Migratory** - Migrating from place to pace.

**Minor pests** - Insect or non-insect pests that cause a loss ranging from 5.0 - 10.0 per cent

**Monophagous pests** - Pests that infest only a particular species of plants alone.

**Mycoplasma** - Mycoplasmas are the smallest, self-replicating organisms belonging to Genus mycoplasma of the Family Mycoplasmataceae They are pleomorphic gram negative, chiefly non-motile bacteria that lack a cell wall.

**N**

**Naphthalene** - Coal tar derivative used as a fumigant against clothes moths, also employed as a soil fumigant, relatively non-toxic to mammals

**Narrow spectrum insecticide** - Insecticide effective against only a narrow range of insects.

**Nematicide** - A chemical compound or physical agent that kills or inhibits nematodes.

**Nerve poison** - Chemical formulations associated with the solubility of tissue lipid and actively block acetyl cholinesterase in warm-blooded animals; Chemical formulations that affect the nervous system of insects (e.g.) Organochlorine, Organophosphate and Carbamate insecticides.

**Nocturnal** - Active during the night time.

**Nuclear polyhedrosis virus (NPV)** - A virus that causes viral disease of insects, mainly the larvae of certain Lepidoptera and Hymenoptera, characterized by the formation of polyhedral inclusion bodies in the nuclei of the infected cells.

**O**

**Obligate parasite** - A parasite that can lead a parasitic mode of life and which cannot exist in any other way.

**Occasional pest** - Pest that reaches significant levels only occasionally (e.g.) Rice case worm.

**P**

**Parasite** - An organism deriving all or a portion of its nutrition from another organism (the host), while permanently or temporarily attached to or live within the host.

**Parasitoid** - Entomophagous insects, which live in or on the body of another insect (host) or on any stage of the host from which they get protection and nourishment at least during any stage of their life cycle.

**Paris green** - An arsenical insecticide, chemically copper acetoarsenite, highly toxic to mammals.

**Permanent pests** - Pests, which have always a foothold on the host whether they have a long or short life cycle and they may have a resting period spent on the host or in the soil (e.g.) Mango hopper, coccids etc.; Also called as persistent pests.

**Persistence** - The property of a pesticide to remain effective as a residue; The property of a pesticide to remain active for long period after application.

**Pest** - Pests are defined as those organisms, which compete with man in his food supply, damage his possessions and attack his person; Pests include insects, mites, ticks, nematodes, fungi, bacteria, viruses, mycoplasmas, protozoa, weeds, rodents, birds, molluscs, crustaceans etc.

**Pesticide** - A chemical formulation by virtue of its toxicity (poisonous properties) that is used to kill pest organisms. The term pesticide includes insecticides, acaricides, nematicides, fungicides, bactericides, herbicides, molluscicides, rodenticides etc.

**Pesticide residue** - Pesticide remaining on or in a plant, plant product, air, water or treated area following a time lapse after application.

**Pest resurgence** - The rapid increase in numbers of a pest following cessation of control measures or resulting from development of resistance of a particular pest to certain plant protection chemicals or due to elimination of its natural enemies as a result of plant protection measures.

**Pest surveillance** - A systematic monitoring of the pest population, dispersion and dynamics in different crop growth phases, so as to forewarn the farmers to take up timely crop protection measures needed.

**Phanerogamic parasite** - A phanerogamic plant is a flowering parasitic plant that derives some or all of its nutritional requirements from another living plant.

**Pheromones** - Chemical substance that attracts members of the same species or one sex of that species, specially insects.

**Physical control** - Control of pests by physical means such as, heat, cold, electricity, sound waves etc.

**Phytoplasma** - Phytoplasmas are obligate, intracellular, cell wall less parasites of plant phloem tissue belonging to a group of bacteria that are related to mycoplasmas.

**Phytosanitation** - Measures requiring the removal or destruction of infested or infected plants or plant parts, which are likely to form a source of reinfestation or reinfection.

**Phytotoxicity** - Substances poisonous or toxic to plants; Substances or chemical compounds or formulations such as, insecticides, fungicides, nematicides etc., which may cause adverse side effects (harmful effects) to the crop.

**Plasma trasfusion** - Plasma is the liquid part of the body's blood. Plasma tranfusion is frequently used to treat critically ill patients.

**Pneumatic sprayer** - It is an spraying equipment that throws the drops of the spray fluid over the plant surface by the action of pneumatic power that is characterized by an extremely fast speed of propelled air, which creates such small droplets like a mist.

**Poison bait** - An attractant foodstuff for insects, molluscs or rodents, mixed with an appropriate toxicant to attract and kill the target organisms.

**Polyphagous pests** - Pests infesting more than one unrelated species of plants

**Positive phototrophic** or **Photopositive** - Natural instinct that makes organisms to move towards artificial light source. **Predator** - An organism that lives by preying upon other living organisms.

**Proprietary name** - Distinguishing name given by the manufacturer to his particular formulated product; Trade name.

**Pterygotes** - Insects, which have wings; Winged insects.

**Q**

**Quarantine** - All operations associated with the prevention of importation of unwanted organisms into a territory or their exportation from it; Regulation forbidding sale or shipment of plants or plant parts, usually to prevent disease, insects, nematodes or weed invasion in a new area.

**Quiescent** - A state of relative biological inactivity induced by physical factors such as, low temperature or other adverse environmental conditions.

**R**

**Random pests** - Casual visitors such as, weevils, cockchafer beetles, blister beetles etc., which have no seasonal basis.

**Recurring pests** - Pests infesting specific seasonal crops such as, pulses, vegetables, grasses etc. regularly. They may pass through several generations during the growth period of the respective crop, rest for a while on alternate hosts and transfer their attention again to the second season crop (e.g.) Rice stem borer, lady bird beetles etc. They are also known as regular pests.

**Resistance to pest outbreak** - The inherent morphological, physiological or genetic properties present in a particular variety of the host that deters infestation by a pest.

**Resistance to insecticide** - The ability of some strains of insects to survive normally lethal doses of an insecticide; The ability having resulted from survival of tolerant individuals in populations exposed to the toxicant for several generations.

**Respiratory poisons** - Chemicals, which block cellular respiration as in the case of fumigants (e.g.) Hydrogen cyanide, carbon monoxide etc.

**Resurgence** - An increase of a pest population after a period of decline to a level higher than its original one, especially when the decline is caused by a pest control measure.

**S**

**Sedentary** - Stationary without any bodily movement; Remaining in a fixed location.

**Seasonal pests** - Pets that occur during a particular season on a particular crop (e.g.) Red hairy caterpillar on groundnut.

**Selective insecticide** - An insecticide, which kills selected insects but spares many or most of the other organisms including beneficial species either through differential toxic action or the manner in which the insecticide is used.

**Semilooper** - A caterpillar in which only one or two pairs of the abdominal legs are wanting and when in movement only small loops are formed (e.g.) Castor semilooper

**Semio chemicals** - Substances that modify the behavior of insects by inducing in them the response to sex, feeding, fear, alarm etc.; These are further classified into pheromones, secreted by insects for stimulating the insects of the same species (intra specific stimulators) and allelo chemicals, the substances used for interspecific communication.

**Severe pest** - A pest with a general equilibrium position above the economic injury level, which makes the pest a constant problem.

**Sex pheromone** - The substance generally released by he females to attract the males for the purpose of mating.

**Sex lure** - A chemical substance that attracts only the members of one sex of that species, especially in insects.

**Siglure** - A synthetic attractant for the Mediterranean fruit fly.

**Soil fumigant** - A pesticide that will evaporate quickly under normal temperature, which when incorporated into the soil, releases a toxic gas that kills pests in the soil.

**Solar light traps** - Solar light trap is a devide for pest control. The device gets charged in the day time using sunlight and automatically switches on at dawn and dusk to trap harmful insects active during night.

**Sporadic pests** - Pests that infest certain crops occasionally within some limited areas or pockets in a more or less severe form (e.g.) Rice earhead bug

**Sterile male technique** - A method of pest control, which involves the release of large numbers of artificially sterilized male insects in the populations.

**Stubble** - Crop residue left over in the field after the harvest of the crop.

**Surfactant** - Surfactant is a substance when added to a liquid reduces its surface tension thereby increases its spreading and wetting properties.

**Suspension** - A state in which solid particles are kept dispersed in a liquid.

**Swarms** - A large number of insects moving together simultaneously.

**Symptom** - A visible, abnormal change in a host (including behavior) as a result of pest infestation.

**Synergists** - Synergists are chemicals that make insecticides more effective at killing pests, but have very little impact on the insects on their own. Piperonyl butoxide is such a common synergist.

**Systemic insecticide** - An insecticide, which is absorbed through a plant surface and is translocated away from the site of application.

**T**

**Typical host** - A host in which pathogenic microorganism or parasite is commonly found; Synonymous with 'natural host'.

**Toxicity** - The capacity of a compound to produce injury; Ability to poison or to interfere adversely with vital processes of the organism by physico-chemical means.

**Translocation** - Transfer of nutrients or viruses throughout the plant; Movement of chemical substances within a plant.

**Trap crop** - A crop, which is preferred by a particular pest in addition to the main crop, but is less important or less valuable than the main crop.

**V**

**Vector** - A living organism such as, insect, mite, bird, higher animal, nematode, parasitic plant and man able to carry and transmit bacteria, viruses, mycoplasma etc.

**Vertical resistance** - A type of resistance effective against certain specific biotypes of the insect, but not against others; Also called as specific resistance.

**Volatile compound** - A compound, which evaporates or vaporizes (changes from a liquid to a gaseous state) at ordinary temperatures or on exposure to air.

**W**

**Water soluble powders and concentrates (WSP and WSC)** - Insecticidal powders or liquids that will go into solution when mixed with water for dilution before spraying without being dispersed as in the case of wettable powders or emulsifiable concentrates

**Wettable powders (WP)** - Insecticide formulations consisting of finely powdered solid particles of active ingredient and inert carriers, which form a suspension when mixed with water for spraying.

# Color Plates

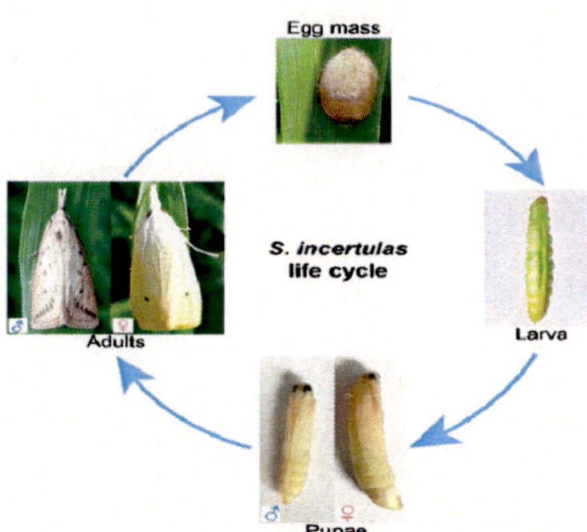

Major pest -Rice stem borer

Minor pest- Red cotton bug

Regular pest - Groundnut leaf miner

Seasoned pest - Castor semilooper

Occasional pest
- Castor slug caterpillar

Persistent pest
- Cardomum thrips

Sporadic pest
-Rice earhead bug

Potential pest - Rice leaf folder

Migratory pest - Desert locust

'r' pest - American cotton boll worm

Intermediate pest - Castor hairy caterpillar

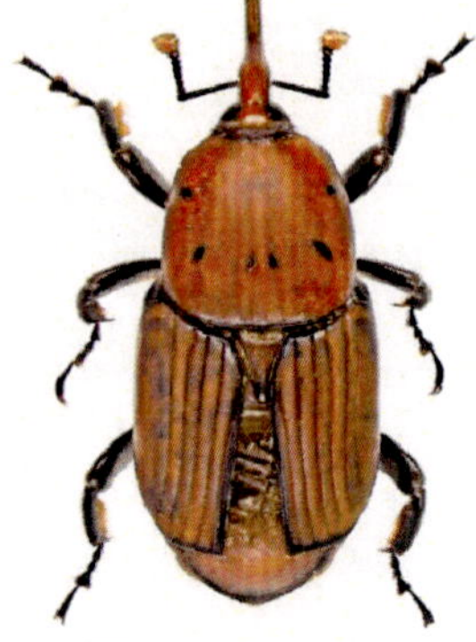
'K' pest -Coconut Red palm weevil

Epidemic pest - Cutworm

Pneumatic hand sprayer

Hydraulic Knapsack sprayer

Foot sprayer

Rocker sprayer

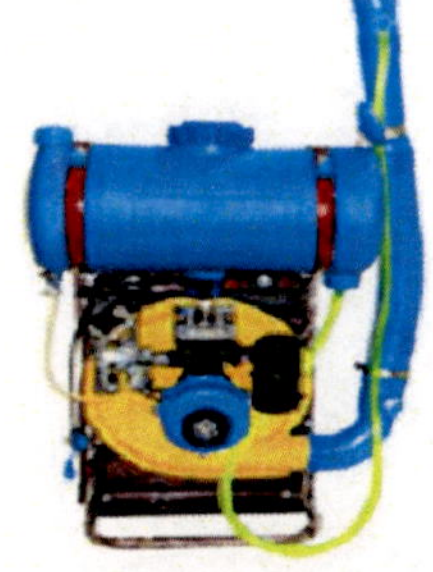

Power mist sprayer

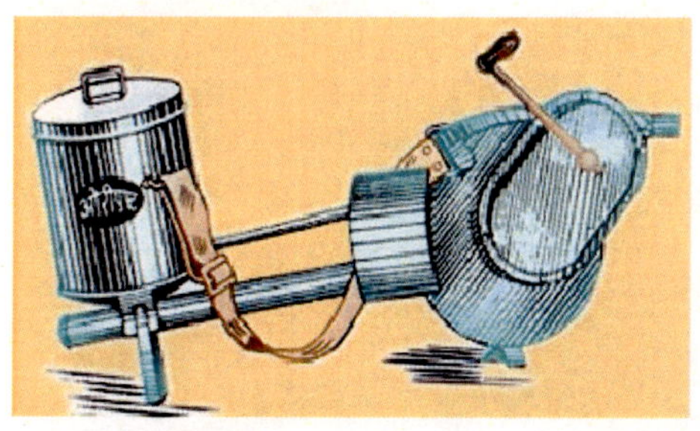

Hand rotary duster

Hand spraying

Tractor mounted sprayer

Aerial spraying

Bird scarer

Bird scarer

Box type rat trap

Jaw type rat trap

# Index